Calculus in Space– Leaving the Plane

by Alan Lipp

Publisher: Tom Maksym

Executive Editor: Steven Jay Griffel

Editor: Mark Weinfeld

CopyEditor: Pat Smith

Proofreader: Dee Josephson

Vice President of Production and Manufacturing: Doreen Smith

Production Manager: Jason Grasso

Project Manager: Matthew Hjembo

Senior Book Coordinator: Paul Zakrzewski

Production Assistants: Richard Lehmbeck, Hugh Haggerty

Designer: Eric Dawson

Technical Art: Alan Lipp, Justin Ruths, Sal Esposito, Matthew Hjembo, Paul Zakrzewski

PEOPLES PUBLISHING GROUP

ISBN 1-4138-1320-8

The Peoples Publishing Group, Inc.
299 Market Street
Saddle Brook, New Jersey 07663

Printed in the United States of America.

10 9 8 7 6 5 4 3 2

Dedication

To my teacher, Mannis Charosh, and my father, Sunny Lipp, who taught me to love mathematics

Acknowledgments

This book would not have been possible without the feedback I have gotten, year after year, from my Calculus students. I also want to acknowledge my editor, Mark Weinfeld, whose encouragement and common sense has helped to greatly improve this book. Any remaining errors, however, are mine alone.

TABLE OF CONTENTS

Introduction

A perennial problem for teachers of Advanced Placement Calculus is what to do after the AP exam. This book provides one answer that I have found particularly successful over the past fifteen years. I spend the last month of the year introducing multivariable calculus. For those students who continue with Calculus Two in college, this introduction offers a solid footing in what is, for many, a treacherous course. Whether students continue on into higher calculus or not, this introduction allows them to take what they have learned in the past eight months and apply it in a fresh and novel setting.

I find that most students are fascinated by the idea that you can take calculus out of the familiar realm of flat graph paper and bring it into space. In addition, software for visualizing three-dimensional surfaces is powerful and stunning. Interest levels soar exponentially when we enter the Math Lab.

This book was specifically written with high school students in mind. But it provides a gentle introduction to multivariable calculus that should be helpful for beginners of all ages who want to get a foundation in the subject before venturing too far into the highly abstract and technical. The style is informal and nonthreatening, and all new concepts are supported with many solved examples.

Students learn best when they can connect new information to prior knowledge. So, throughout the text, you will find suggestions for connecting key three-dimensional ideas such as tangent planes, partial derivatives, and double integrals to the analogous two-dimensional concepts of tangent lines, derivatives, and integrals.

Each chapter is short and limited in scope. The intent is to make each one accessible as a single lesson spanning two days. Each chapter includes a problem set with problems for practicing the skills presented, as well as more thought-provoking questions that extend the ideas into new regions.

Chapter 1 introduces three-dimensional graphing. Some of these ideas may be familiar from earlier courses, but the chapter emphasizes the description of surfaces and regions with multivariable equations and inequalities.

Chapter 2 focuses on multivariable equations and how to visualize the surfaces they describe. This chapter introduces sections of a function as a tool that uses two-dimensional graphs to help develop a three-dimensional image.

Chapter 3 continues the ideas from the previous chapter, but emphasizes level curves. Connections are built between these surface contours and implicit functions, which are studied earlier in calculus.

Chapter 4 develops the idea of partial derivatives as ordinary derivatives within sections. Students have the opportunity to practice differentiating multivariable functions and interpreting these derivatives in terms of the familiar properties of two-dimensional functions.

Chapter 5 examines surfaces formed by revolution. These are used to begin the discussion of volume within a familiar context. This chapter helps students create multivariable functions for these surfaces that, earlier in the year, were only described as solids of revolution. Now these surfaces are seen to have their own multivariable descriptions.

Chapter 6 introduces double integrals and their evaluation as iterated integrals using the familiar method of slabs. Chapter 7 extends the application of double integrals to finding the average value of a multivariable function, a procedure that is similar to finding the average value of a single variable function.

Finally, Chapter 8 offers a final review, consisting of a collection of problems covering the earlier chapters followed by a self-assessment test.

For the past fifteen years, my students have questioned me, stretching me to find new clarity to help them come to understand calculus. They have pushed me off the beaten path to walk uncharted byways with them. It has been a wonderful journey that we, together, have enjoyed taking. I have tried to convey that excitement and fun in the writing of this book. I welcome hearing from you, my readers, about your own journey when leaving the plane and taking calculus into space. Your comments and criticisms of this volume will help me to improve future editions. You can reach me at alipp@williston.com.

Alan Lipp

South Deerfield, Massachusetts

CHAPTER 1

Moving into Space

Imagine a $10 \times 10 \times 10$ cube sliced into 1,000 unit cubes like a giant Rubik's cube™. If the cube sits on the first quadrant of a coordinate plane with one corner at the origin, then we can identify each small cube in the base by the coordinates of its lower left corner. In order to identify the cubes above this level, we would need a third coordinate. The cube that is five levels above the cube at (4, 8), for example, would be at (4, 8, 5). By using three coordinates, (x, y, z), we can identify the position of any of the cubes. In fact, with three coordinates and three axes we can identify any point in space.

Look at one of the corners of the room you are in, where two walls meet the floor at a point. This is what a three-dimensional coordinate system looks like. The axes are conventionally drawn as in Figure 1.1 so that we are looking down on the xy-plane from above the first quadrant.

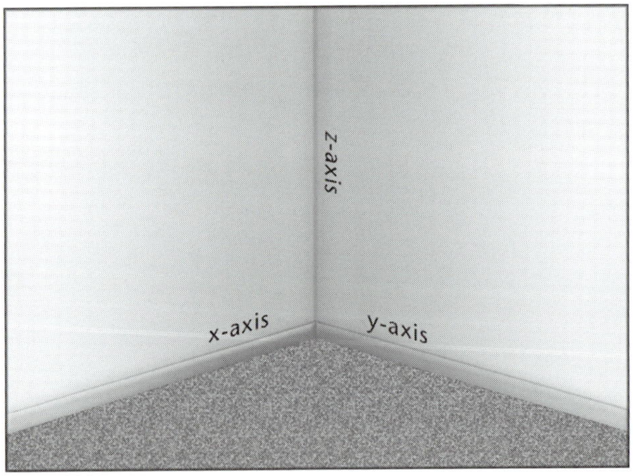

Figure 1.1

The three axes intersect at the point (0, 0, 0), which is called the *origin*. Look again at the corner of your room. The two walls and the floor that meet there are parts of planes. The plane in which the familiar x- and y-axes lie is called the *xy-plane*. The plane including the x- and z-axes is the *xz-plane*, and the plane including the y-axis and the z-axis is the *yz-plane*. Together, these three planes are called the *coordinate planes*. Just as the x- and y-axes divide the xy-plane into four quadrants, the three coordinate planes divide space into eight *octants*.

When we graph in two dimensions, the equations $x = c$ and $y = k$ describe vertical and horizontal lines, respectively. We can also use equations to describe parts of space. What does the equation $x = 5$ mean when we are in three dimensions? This equation is satisfied by any point whose x-coordinate is 5. In two dimensions, $x = 5$ represents a line parallel to the y-axis and 5 units to the right. Now that we are in space, however, we can move 5 units in the positive x-direction away from any point in the yz-plane. These points form a plane parallel to the yz-plane. We can think of this plane as a shift of the yz-plane five units in the positive x- direction.

Example 1 Describe the regions in space defined by the equations **a.** $y = -2$, **b.** $|z| = 3$, and **c.** $x + y = 2$

Solution In two dimensions, $y = -2$ is a horizontal line 2 units below the x-axis. In three dimensions, any point with y-coordinate -2 satisfies the equation and these points all lie on a plane parallel to the xz-plane and 2 units away from it in the negative y direction.

The equation $|z| = 3$ is satisfied by both $z = 3$ and $z = -3$ so this equation describes two planes, each parallel to the xy-plane. One of the planes is 3 units above the xy-plane and the other is 3 units below it.

In the xy-plane, the equation $x + y = 2$ graphs as a line with intercepts (2, 0) and (0, 2). The sum of the x- and y-coordinates of any point along this line will be equal to 2. In three dimensions, this equation is still satisfied by the line $x + y = 2$ when $z = 0$. Since the equation places no restriction on z, however, it also holds when $z = 7$, $z = -2$, or, in fact, for any value of z. The collection of all of these lines is a plane, perpendicular to the xy-plane; it is shown in Figure 1.2. The plane $x + y = 2$ intersects the xy-plane in the two-dimensional line $x + y = 2$. It intersects the xz-plane (where $y = 0$) in the line $x = 2$, and intersects the yz-plane in the line $y = 2$.

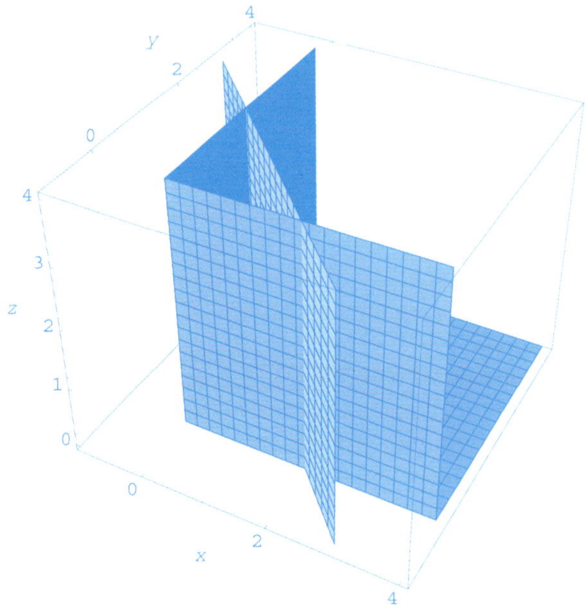

Figure 1.2

By the way, Figure 1.2 illustrates the use of a surrounding box to display a scale and to lend perspective to the drawing. The box itself is not part of the graph.

Free Variables

When we graph the equation $y = -16$ in two dimensions we get a line because x can take on any value. A variable that is missing from an equation is called a *free variable* because we are free to let that variable take on any value. The equations in the previous example each had one or two free variables.

When we are working in three dimensions, we have the possibility of writing equations with zero, one, or two free variables. When two variables are free, as in $x = 11$ or $z = -18$, the equation is a plane parallel to one of the coordinate planes. The graph of $x + y = 2$, as discussed in Example 1, illustrates what can happen when there is just one free variable.

The next example further illustrates what can happen when there is just one free variable in an equation.

Example 2 Describe the points in space that satisfy the equation $y^2 + z^2 = 100$.

Solution If we did not know that this was supposed to be a three-dimensional graph, we might think of this as a circle of radius 10 in the yz-plane. In three dimensions, however, that is just the part of the surface that lies in the plane $x = 0$. Since x is a free variable, however, we get exactly the same equation with the same graph in any plane parallel to the yz-plane. The collection of these parallel circles forms an infinitely long tube of radius 10 surrounding the x-axis. The three-dimensional graph, thus, is the surface of a cylinder and the parallel circles are cross-sections of the cylinder with any plane $x = c$.

Figure 1.3 shows a portion of this cylinder. You can visualize this cylinder as the surface formed by shifting the circle $y^2 + z^2 = 100$ in the x-direction.

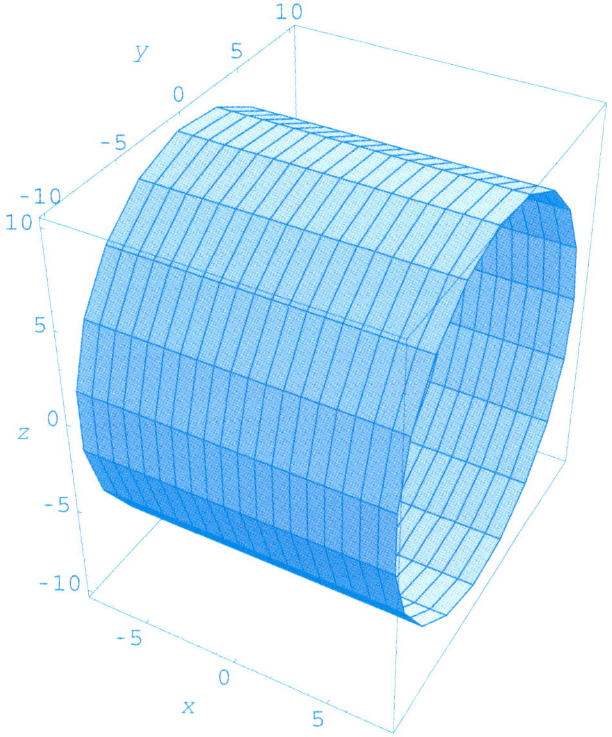

Figure 1.3

Multivariable Equations

In algebra and calculus we graph equations with two variables where, typically, x is the independent variable and y is the dependent variable. If the equation represents a function, we write $y = f(x)$. A *multivariable equation* can have any number of independent variables. In this book, however, we will consider only equations with at most two independent variables since these can be visualized as surfaces in space.

When we graph in three dimensions, both x and y are taken as independent variables and z is considered the dependent variable. A two-variable function $z = g(x, y)$ determines one and only one z-value for each choice of x and y in the domain. The domain of a two-variable function can be any region in the plane; however, it is often chosen to be a rectangle $[x_{min}, x_{max}] \times [y_{min}, y_{max}]$. Either or both of these intervals can be open rather than closed and either or both can be infinite in extent.

A useful way to think of "each choice of x and y" is to imagine choosing any point (x, y) in the xy-plane as input and, for each such point, computing a value of z. We can visualize these values as points (x, y, z), which are z units above or below the input point (x, y). The collection of all such points, for all possible inputs (x, y), forms a surface in space.

In the plane, a function must pass the *vertical line test*, that is, any vertical line $x = c$ can only intersect the curve in at most one point. The corresponding test for functions in space is that each line through (x, y), perpendicular to the xy-plane, can only intersect the surface in one and only one point.

Space Cylinders

Example 2 showed that the equation of an ordinary two-dimensional circle describes the surface of an infinitely long cylinder in space. Something similar happens when any equation is taken out of the plane and into space. An equation that graphs as a curve in the xy-plane, when treated as a multivariable equation with a free variable, graphs as an identical curve in every plane parallel to the xy-plane. Together, these curves form surfaces in space called *cylinders*. Note that unlike the most familiar cylinders, these surfaces are infinite in extent and do not have to be closed; the surface $y = x^2$, for example, is called a *parabolic cylinder*.

In the same way, an equation where y is a free variable graphs as a curve in the xz-plane and as a space cylinder perpendicular to the xz-plane. An equation with the free variable x will graph as a cylindrical surface perpendicular to the yz-plane.

Example 3 Describe the following surfaces

a. $\frac{x^2}{4} + \frac{(y+3)^2}{9} = 1$ **b.** $\frac{x^2}{25} - \frac{z^2}{49} = 1$ **c.** $z = -(y-4)^2 + 3$ **d.** $z = 2\sin(3x)$

Solution In two dimensions, the equation in **a.** is an ellipse centered at $(0, -3)$ with axes of length 2 and 3. In three dimensions, since z is a free variable, this is an *elliptical cylinder*. Figure 1.4a shows that this cylinder is perpendicular to the xy-plane.

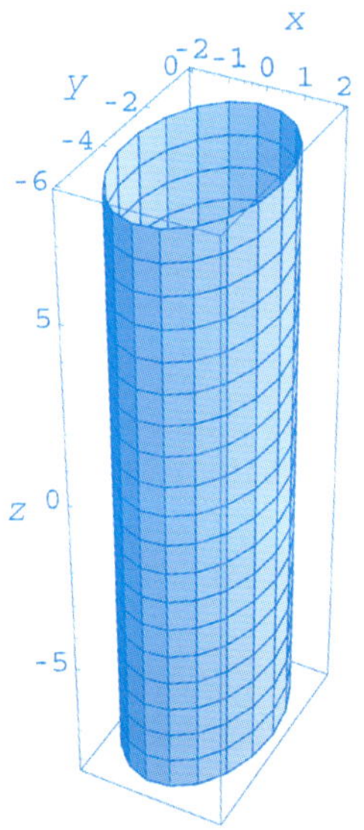

Figure 1.4a

Equation **b.** has y as a free variable. In two dimensions, it graphs as a hyperbola in the xz-plane, centered at the origin, and asymptotic to the lines $\frac{x}{5} = \pm\frac{z}{7}$. Figure 1.4b shows this surface, a *hyperbolic cylinder* perpendicular to the xz-plane. Each plane $y = c$ intersects the surface in an identical hyperbola. The collection of all these hyperbolas forms the cylinder that, in space, is asymptotic to the *planes* $\frac{x}{5} = \pm\frac{z}{7}$.

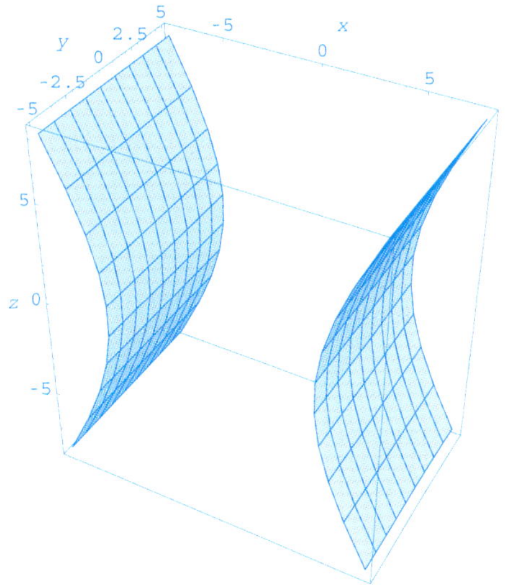

Figure 1.4b

Since x is a free variable, the surface in **c.** intersects each plane $x = c$ in the parabola $z = -(y - 4)^2 + 3$, which has its vertex at (4, 3) in the yz-plane and has $y = 4$ as its axis of symmetry. These emerge from the plane as a *parabolic cylinder* where $y = 4$ is the *plane of symmetry*, as shown in Figure 1.4c.

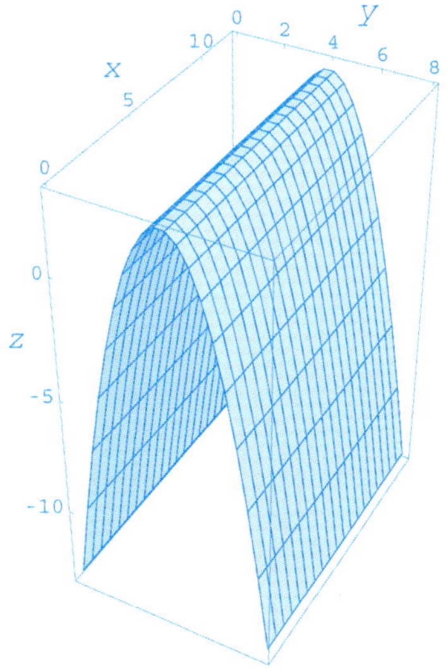

Figure 1.4c

Finally, $z = 2\sin(3x)$ has y as a free variable. In two dimensions, it graphs as a sinusoid in the xz-plane and, in three dimensions, as a *sinusoidal cylinder*. The two-dimensional curve has period $\frac{2\pi}{3}$ and amplitude 2. These properties extend into space where the cylinder varies periodically in the x-direction with period $\frac{2\pi}{3}$. The z-values of the surface satisfy $|z| \leq 2$, so the surface lies between the parallel planes $z = -2$ and $z = 2$. This cylinder is shown in Figure 1.4d.

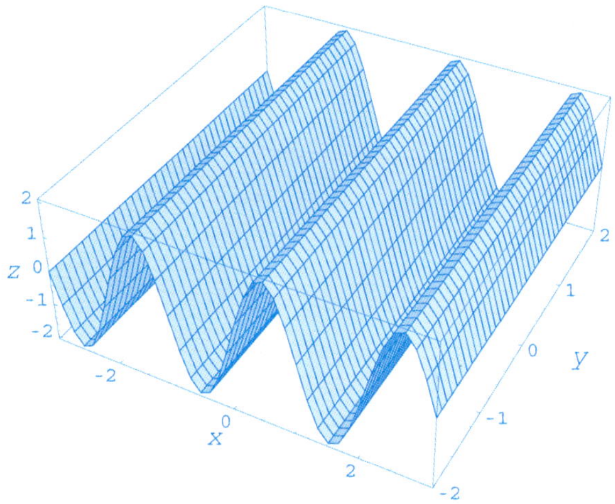

Figure 1.4d

Distance in Space

The distance between any two points in the plane can be computed using the distance formula $D = \sqrt{(x_2 - x_1)^2 + (y_2 - y_1)^2}$. This formula is just the Pythagorean theorem applied to the triangle whose hypotenuse is the segment joining the points (x_1, y_1) and (x_2, y_2) and whose legs are vertical and horizontal segments through these two points. There is a similar formula for the distance between any two points in space (x_1, y_1, z_1) and (x_2, y_2, z_2). This formula is also derived from the Pythagorean theorem. Figure 1.5 illustrates the process.

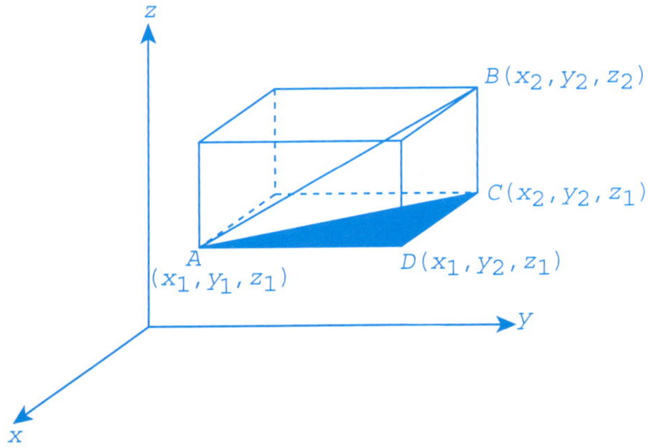

Figure 1.5

The points $A(x_1, y_1, z_1)$ and $B(x_2, y_2, z_2)$ lie at opposite corners of a box in space. The distance formula enables us to compute the length of the diagonal in space through the interior of the box, joining these points. Here, $\overline{AB}$ is the hypotenuse of $\triangle ABC$ where C has coordinates (x_2, y_2, z_1) and lies in the plane of the bottom of the box, which is $z = z_1$.

If we draw $\overline{AC}$, we form another right triangle, $\triangle ACD$. Since this triangle lies in the plane $z = z_1$, we can compute the length AC using the ordinary two-dimensional distance formula. The legs of $\triangle ABC$, thus, are $BC = z_2 - z_1$ and

$AC = \sqrt{(x_2 - x_1)^2 + (y_2 - y_1)^2}$. Now you can apply the Pythagorean theorem to $\triangle ABC$

to get $AB^2 = (x_2 - x_1)^2 + (y_2 - y_1)^2 + (z_2 - z_1)^2$ so that

$AB = \sqrt{(x_2 - x_1)^2 + (y_2 - y_1)^2 + (z_2 - z_1)^2}$.

Example 4 Find the length of the median of $\triangle ABC$ shown in Figure 1.6.

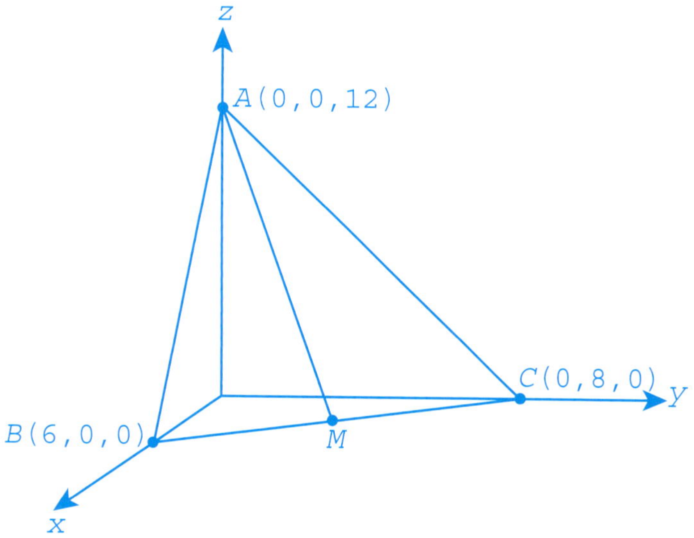

Figure 1.6

Solution The coordinates of the midpoint are of $\overline{BC}$ are (3, 4, 0), the average of the coordinates of B and C. Apply the distance formula to $A(0, 0, 12)$ and $M(3, 4, 0)$ to get that $AM = 13$.

Space Regions and Inequalities

Just as we can describe regions in the plane with inequalities, we can use inequalities to describe regions in space.

Example 5 Describe each of the regions in space given by the following inequalities.

a. $x > 0$ **b.** $|y| < 4$ **c.** $(x-2)^2 + y^2 + (z+8)^2 > 36$ **d.** $y > 0$ and $z < 0$

Solution The region in **a.** contains all points whose x-coordinate is positive. These points lie to the positive x side of the yz-plane, so this inequality describes the "half-space" where x is positive.

The region in **b.** contains the points whose y value satisfies $-4 < y < 4$. This is the inside of a sandwich where the "bread" is the parallel planes $y = 4$ and $y = -4$. Points that lie between but not on these two planes are included in this region.

The inequality in **c.** is the same as $\sqrt{(x-2)^2 + y^2 + (z+8)^2} > 6$. We can recognize the left-hand side as the distance between two points and use this idea to describe the region defined by this inequality. It includes all points (x, y, z) that are more than 6 units away from the point $(2, 0, -8)$. That is, this inequality describes the points outside a sphere of radius 6 centered at $(2, 0, -8)$.

Finally, in **d.**, if $y > 0$ and $z < 0$ then we must be in front of the xz-plane (where $y > 0$) and below the xy-plane (where $z < 0$). Therefore, these inequalities describe two of the eight octants.

Problems

1. How far is the point $(3, -5, 2)$ from each of the three coordinate planes?

2. Name three points in the plane $z = 2$.

3. Find the distance between the centers of the spheres $(x-2)^2 + (y-5)^2 + (z+10)^2 = 9$ and $(x-7)^2 + (y-1)^2 + (z+7)^2 = 1$.

4. What is the equation of the plane formed by shifting the xz-plane parallel to itself and 4 units in the positive y-direction?

5. What is the equation of the plane containing $(4, -2, 0)$, $(5, -10, 0)$ and $(1, -2, 0)$?

6. Name three points between the plane $z = 2$ and the xy-plane.

7. Consider the cylinder $x^2 + y^2 = 25$.
   a. What is the central axis of the cylinder?
   b. Find any two points on the cylinder that do not lie on the circle $x^2 + y^2 = 25$ in the xy-plane.

   Any point P on the cylinder lies on a circle and on a line lying on the surface of the cylinder. Just think of cutting a cylinder perpendicular to its central axis (like slicing a carrot) or cutting the cylinder through the central axis (like splitting a log).

   c. If the coordinates of P are (a, b, c), give the coordinates of three other points on the circle containing P.
   d. If the coordinates of P are (a, b, c), give the coordinates of two other points on the line containing P.

8. Name three points on the plane $y = 2x - 5$.

9. How far is the point $(3, -5, 2)$ from each of the coordinate axes?

10. Find the equation of a plane containing the points $(4, 3, 3)$, $(7, 2, 4)$, and $(-8, 1, 5)$. Hint: Focus on just the y- and z-coordinates.

11. Find a system of inequalities describing the interior of a $4 \times 4 \times 4$ cube centered at the origin.

12. **a.** Find the coordinates of the other four corners of a rectangular box if four of its corners are $(4, 6, 12)$, $(4, 1, 12)$, $(4, 6, 1)$, and $(9, 6, 12)$.
    b. What is the surface area of the box?
    c. What is the volume of the box?

13. Find the equation of a sphere centered at $(3, -8, 4)$ and tangent to the nearest coordinate plane.

14. **a.** Find the equation of two planes that are tangent to the sphere $x^2 + (y-3)^2 + (z+2)^2 = 25$ and parallel to the xy-plane.
    b. Find the equations of two tangent planes that are parallel to the yz-plane and of two tangent planes that are parallel to the xz-plane.
    c. Find a set of inequalities describing the region outside the sphere but inside the box formed by the six planes determined in **a.** and **b.**

15. **a.** Sketch the solid enclosed by the planes $x = 0$, $y = 0$, $y + 2x = 4$, $z = -2$, and $z = 5$.

b. What is the common name for this shape?

c. Find its volume.

16. **a.** Describe the shape of the solid given by the planes $|x| \leq 2$, $|y - 3| \leq 4$, and $|z - 5| \leq 5$.

b. Find the coordinates of its corners.

c. Find the volume of the solid.

d. How would you change your description in **a.** if the inequalities had been $|x| < 2$, $|y - 3| < 4$, and $|z - 5| < 5$?

17. **a.** Find the volume of the region

$$x^2 + (y - 1)^2 + (z - 2)^2 \leq 49$$

b. Find the area of the intersection of this region with the plane $y = 1$.

18. The equation $x - 2y + 6z = 20$ has no free variables. It describes a plane that is not perpendicular to any of the coordinate planes.

a. What are the coordinates of the three intercepts of the plane with the coordinate axes?

b. This plane and the three coordinate planes enclose a volume in the shape of a triangular pyramid. The volume of a pyramid can be computed from the formula $V = \frac{1}{3}Ah$ where A is the area of the base and h is its height. Find the volume of this pyramid.

19. The spheres $(x + 8)^2 + y^2 + (z + 3)^2 = 16$ and $(x + 8)^2 + (y + 5)^2 + (z + 3)^2 = 9$ intersect in a circle. Find the equation of that circle and its area.

20. Find the equation of a cylinder that is perpendicular to the xz-plane and has a radius of 1.

21. Find a pair of equations for the two cylinders with radius 1 shown in Figure 1.7. They are tangent to each other and to the xz-plane. One of the cylinders is also tangent to the yz-plane.

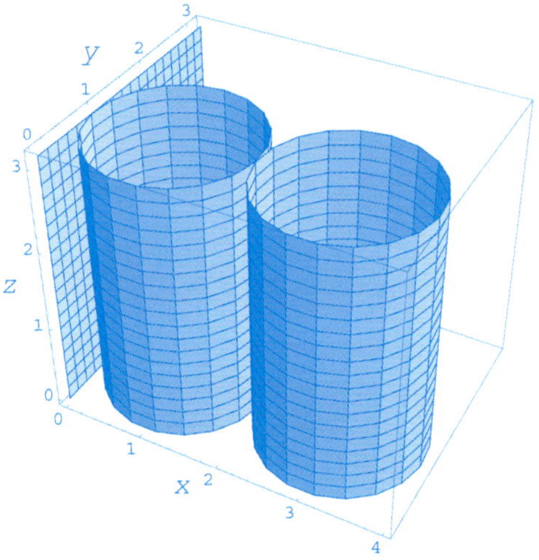

Figure 1.7

22. **a.** Find the coordinates of the points where the plane $y = 3 - 2x$ intercepts the coordinate axes.

b. Find the angle between the plane and the xy-plane. Explain your reasoning.

23. The plane $z + 2y + 3x = 6$ intersects each of the coordinate planes in a line as shown in Figure 1.8. Find the equation of each of these three lines of intersection.

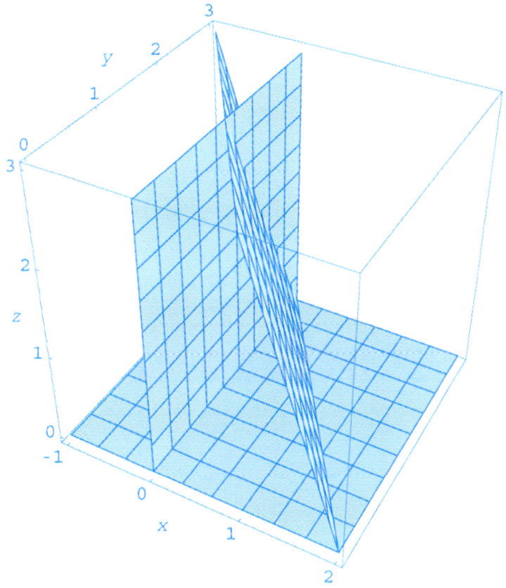

Figure 1.8

24. Find the equation of a sphere centered at $S\,(2, 2, 2)$ and tangent to the three coordinate planes.

25. There is a gap between the sphere in problem 24 and the origin. (Imagine rolling a basketball into the corner of a room. There will be a small space between the corner of the room and the basketball.) Find the equation of a new sphere that will fit into that space and be tangent to the three coordinate planes and to the larger sphere. Figure 1.9 shows the two tangent spheres.

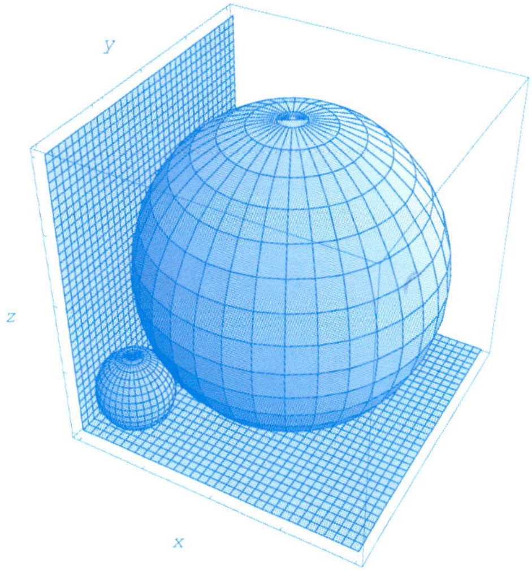

Figure 1.9

Hint: The center of the new sphere **C** lies on the line joining (0, 0, 0) and (2, 2, 2) and so will have the form (c, c, c). The point of tangency of the two spheres T will also lie on this line; call it (t, t, t). Use the distance formula in space to compute the lengths CT and TS. Eliminate t from these equations and solve for c.

CHAPTER 2

Visualizing Surfaces with Sections

In the previous chapter we saw how to extend graphing from two dimensions to three. The graph of an ordinary two-dimensional function is a collection of points $(x, f(x))$ that pairs the input of a function with its output. A *multivariable function*, such as $g(x, y) = x + 2y$, has two input values. Its graph is a collection of points $(x, y, g(x, y))$ that typically forms a surface in space. Two-dimensional graphs are enormously helpful in visualizing the properties of single variable functions; three-dimensional graphs are just as useful in coming to understand the properties of multivariable functions.

The catch, however, is that we cannot easily visualize a three-dimensional figure in a two-dimensional picture. In this chapter, however, we will examine several special types of two-dimensional graphs called *sections* that are very useful in helping us to visualize the surface described by a multivariable function.

It may come as a surprise to you that you have been working with multivariable functions for years, without actually realizing it. Here are a few familiar situations viewed from a multivariable viewpoint.

- The perimeter of a rectangle is a two-variable function, $P(L, W) = 2L + 2W$.
- The volume of a box is a three-variable function, $V(L, W, H) = LWH$.
- The hypotenuse of a right triangle is a two-variable function, $h(a, b) = \sqrt{a^2 + b^2}$.
- The average of five numbers, $A(n_1, n_2, n_3, n_4, n_5) = \frac{n_1+n_2+n_3+n_4+n_5}{5}$, is a five-variable function.

- The area under the parabola $y = x^2$ from $x = a$ to $x = b$ is a two-variable function, $A(a,b) = \int_a^b x^2 dx$.
- If you have taken physics, you may be familiar with the two-variable function for computing the kinetic energy of an object, $E(m, v) = \frac{1}{2}mv^2$.

Each of these and countless other formulas can be viewed as multivariable functions. Most of the previous examples are functions of two variables, but several have more than two inputs. The concept of a multivariable function places no restriction on the number of inputs. However, in this book we will focus our attention on functions of just two variables as these are the simplest to visualize.

Of course, some functions of two variables are easier to visualize than others. Consider, for example, the surface given by the equation $z = \sqrt{100 - x^2 - y^2}$. Since $x^2 + y^2 + z^2 = 100$ is a sphere of radius 10, the given surface is the top hemisphere of radius 10. You probably don't need much help visualizing a hemisphere. However, let's use this example to develop a concept that can be used to help visualize more complex surfaces.

Imagine slicing an orange or a tomato. Each slice is a disk, and the disks vary in size depending upon where they sit in the solid. Now imagine instead a hemispherical surface (for example, just the skin of the top half of the orange). Slice again. This time the slices are curves—semicircles of varying size. These semicircles are called *sections* of the hemispherical surface. The concept of the sections of a hemispherical surface is developed more mathematically in the example below.

Example 1 What are the x-sections and y-sections of the surface $z = \sqrt{100 - x^2 - y^2}$?

Solution Once again, the given surface is the top hemisphere of radius 10. The x-sections are formed by slicing the surface parallel to the yz-plane. If you imagine the hemisphere as an orange skin and the plane $x = 5$ as the knife, then the cross-section is the semi-circle $z = \sqrt{75 - y^2}$. Any choice of $x = c$ in the domain $-10 \leq c \leq 10$ will produce a semicircle in the plane $x = c$ whose equation is $z = \sqrt{100 - c^2 - y^2}$. Notice that these curves lie in parallel yz-planes. Similarly, the y-sections lie in planes $y = c$, all parallel to the xz-plane, and all semicircles of the form $z = \sqrt{100 - x^2 - c^2}$.

As previously commented, semi-circular cross-sections may not be needed to sharpen your image of a hemisphere. However, most multivariable functions are far more complex than this. Figure 2.1, for example, shows the surface $z = f(x, y) = x^3 + 5x^2 - xy - 4y^2$, a multivariable cubic function. It is so complex that even its graph doesn't tell us much.

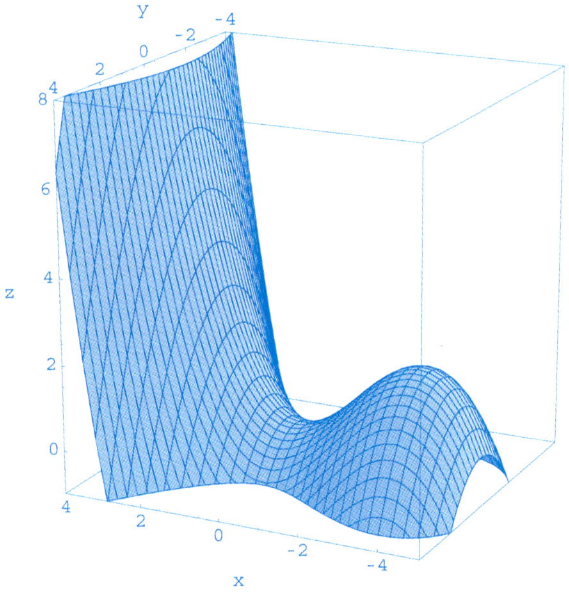

Figure 2.1

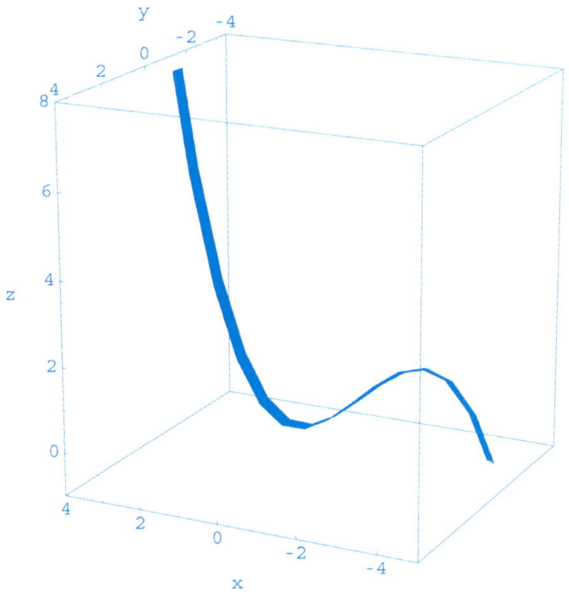

Figure 2.2

In Figure 2.2, we've sliced this surface with the plane $y = -1$ to display its intersection with this plane. Notice that the intersection looks like an ordinary cubic graph.

In fact, if we substitute $y = -1$, we get $f(x, -1) = x^3 + 5x^2 + x - 4$ which *is* an ordinary cubic. There is nothing special about the value -1; any y-value will give a cubic equation in x. The functions $f(x, c)$ are all ordinary cubic equations whatever value of c we choose. Several are shown in Figure 2.3. The intersections of a surface with different planes, as shown in the figure, can give insights into the shape of the surface and the properties of the function.

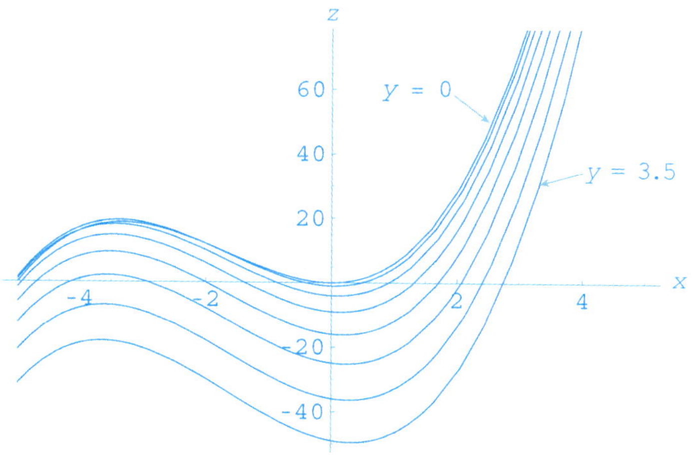

Figure 2.3

Sections of a Surface

A section of a three-dimensional surface is its intersection with a plane parallel to one of the coordinate axes. Intersections with planes parallel to $y = 0$ are called y-sections and intersections with planes parallel to $x = 0$ are called x-sections. Intersections parallel to the xy-plane are sometimes called z-sections, but are more commonly called level curves or contours. Contours are distinguished from x- and y-sections for several reasons. First, x and y are independent variables, while z is dependent. Of greater importance is that x- and y-sections of a function are always functions as well, while z-sections are often implicit equations. In this chapter, we focus on what we can learn from the x- and y-sections of a multivariable function. We will examine z-sections in the next chapter.

Since sections are ordinary two-dimensional curves, we can use all the familiar techniques of calculus to uncover their properties. Section graphs can be helpful in analyzing the three-dimensional properties of complex surfaces.

Example 2 Figures 2.4a and 2.4b show x- and y-sections of the surface $z = 9 - x^2 + y^2$. What do these sections say about the shape of the surface?

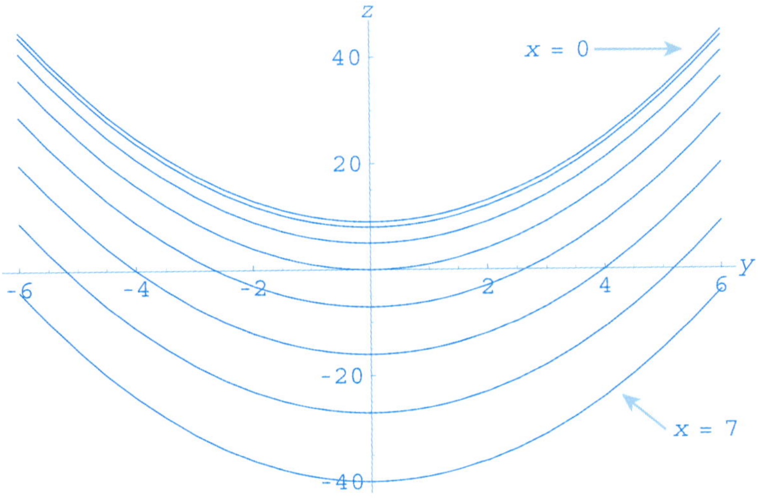

Figure 2.4a

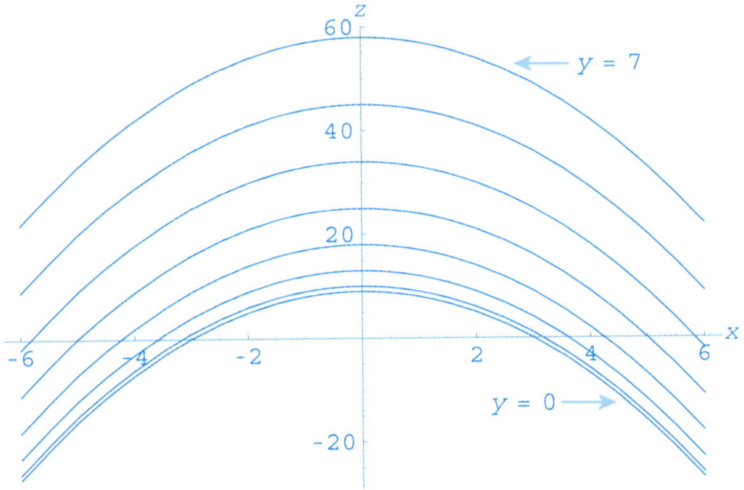

Figure 2.4b

Solution

To find x-sections, choose a sequence of values of x and graph the resulting single-variable functions of y against z. Figure 2.4a shows one set of x-sections of this surface. They are parabolas that open upward, with minima on the z-axis. They belong to the family of parabolas $z = (9 - x^2) + y^2$. For example, if we select $x = 3$, the equation $z = (9 - x^2) + y^2$ becomes $z = y^2$, which is a concave up parabola that turns at (0, 0). This parabola is shown in Figure 2.4a; it is the parabola through the point $y = 0$ and $z = 0$.

Note that, as we increase x, the vertex of the parabola is shifted downward. We can interpret this by imagining that we are standing on the surface $z = 9 - x^2 + y^2$ at some point. If we move in a direction that increases the magnitude of our x-coordinate, then we will move down the surface in the direction of decreasing z values.

If we instead hold y constant, we generate a family of y-sections that are also parabolas. They have the form $z = (9 + y^2) - x^2$ and open downward as shown in Figure 2.4b. As y increases, the z-values of the vertex increase also. If we move along the surface so that $|y|$ increases, then our z-coordinate will increase as well. That is, we move higher up the surface. The idea of putting yourself into the picture and walking on the surface is a very useful technique. It helps to make the abstract ideas of multivariable calculus more concrete and real.

Notice, once again, that x-sections are functions in Example 2 and are all of the form $z = f(y)$. Each section lies in a different x-plane. In order to visualize them, however, it is useful to graph them in the same yz-plane and label them with their x-values as was shown in Figure 2.4a.

Further, all the parabolas are symmetrical across the z-axis. Can you visualize what this suggests? It takes some practice to take these two-dimensional curves and visualize a three-dimensional surface. Perhaps Figures 2.5a, b, and c will help.

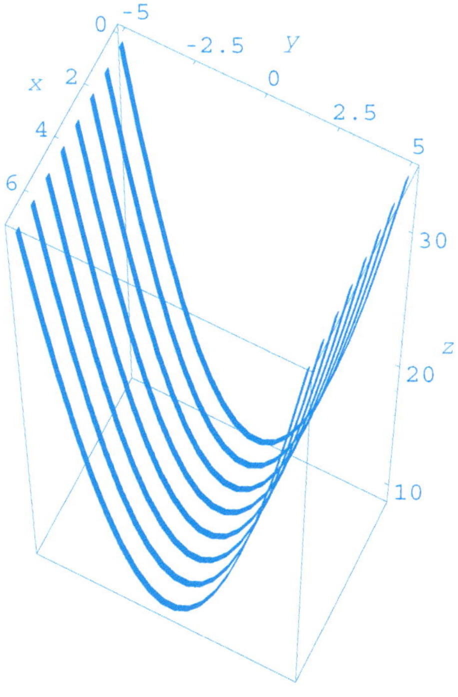

Figure 2.5a

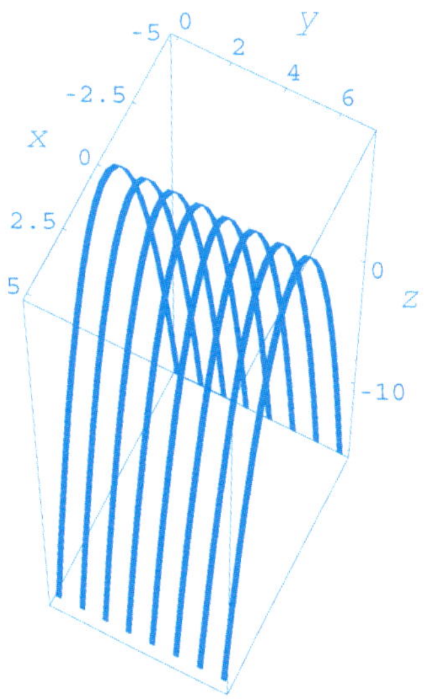

Figure 2.5b

Figure 2.5a shows the same x-sections as Figure 2.4a, but now the sections are arranged in their correct places in space. Likewise, Figure 2.5b is the three-dimensional version of the y-sections in Figure 2.4b.

Together these should help you visualize the saddle-like surface in Figure 2.5c. If you imagine this figure sliced like a loaf of bread, the sections in Figure 2.5a show one set of slices while those in Figure 2.5b show the result of slicing in a perpendicular direction.

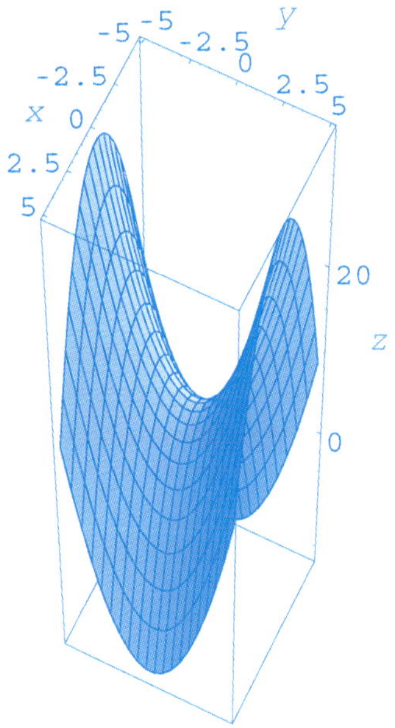

Figure 2.5c

Example 3 What do the x- and y-sections of $z = 9 - x^2 - y^2$ tell us about the shape of the surface?

Solution This is similar to the previous example, but now both the x- and y-sections are parabolas with maxima. The z-axis is the axis of symmetry and the z-values decrease as we move away from the origin in any direction. Add to this fact that the z-sections, $x^2 + y^2 = 9 - z$, are circles centered at the origin with radius $\sqrt{9 - z}$ and you may begin to visualize a surface of revolution. You probably found volumes of solids shaped just like this one in your calculus course. If you rotate the parabola $y = 9 - x^2$ around the y-axis, you get the surface $y = 9 - x^2 - z^2$. The surface $z = 9 - x^2 - y^2$ is identical except that it is rotated around the z-axis instead of the y-axis. There will be more about surfaces of revolution in Chapter 5.

Example 4 What are the x- and y-sections of $f(x, y) = 5\sin(\frac{x}{4}) - 3\cos(\frac{y}{2})$? What do they tell us about the surface?

Solution Remember that x is held constant in x-sections. A typical x-section, therefore, is a cosine wave $z = -3\cos(\frac{y}{2}) + [5\sin(\frac{x}{4})]$. This is a wave whose period is 4π and whose amplitude is 3. As x increases, the wave is shifted vertically (in the z-direction) by $5\sin(\frac{x}{4})$. The vertical shift itself varies sinusoidally. Similarly, the y-sections are sine waves with period 8π and amplitude 5.

The combined effect of these waves is to produce a sinusoidal surface that is periodic in the x-direction with period 8π and periodic in the y-direction with period 4π. If we move along the surface in the direction of increasing x, our z-values vary between a low of -8 and a high of $+8$, the combined effect of the two amplitudes. There is a similar variation of height in the y-direction. The complete surface is illustrated in Figure 2.6.

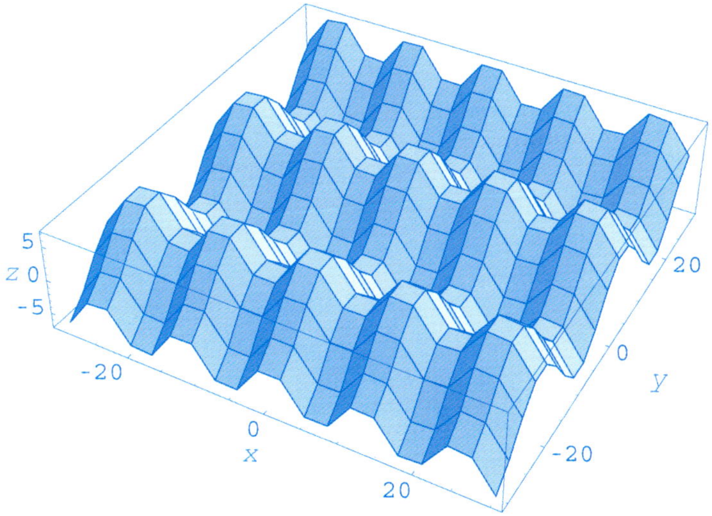

Figure 2.6

Example 5 What surface has the x-sections and y-sections shown in Figures 2.7a and 2.7b, respectively?

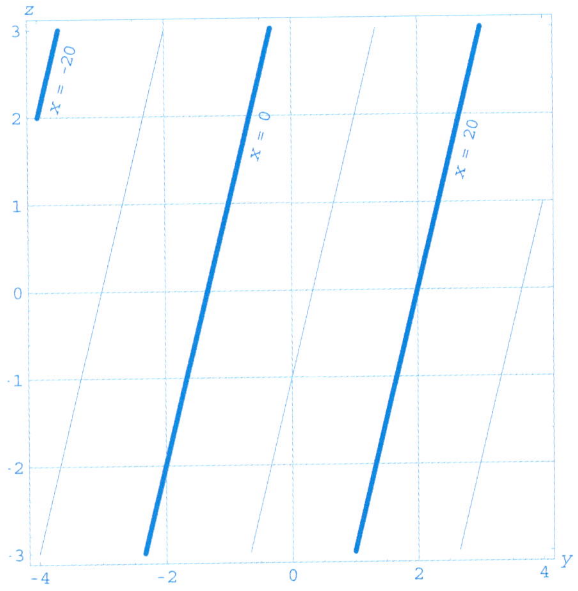

Figure 2.7a

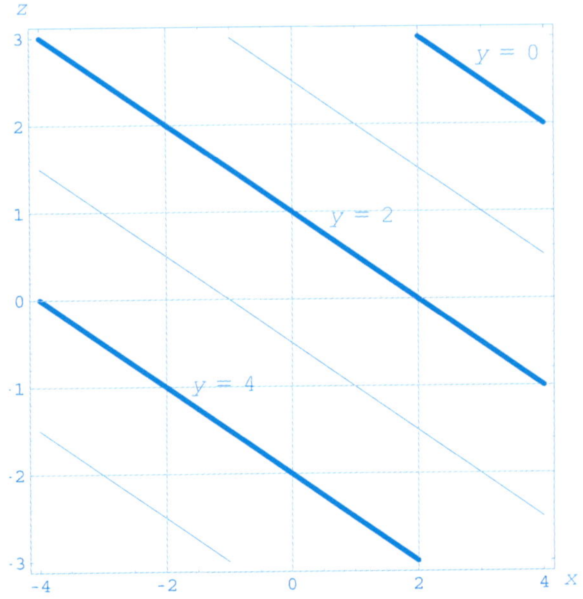

Figure 2.7b

Solution The x-sections are lines with slope 3, so they must have the form $z = 3y + b$. The valus of b vary with the choice of x. The y-sections have slope $-\frac{1}{2}$, so they have the form $z = k - 0.5x$ where k is determined by the choice of y. In other words, z must have the form $z = -0.5x + 3y + c$. To determine c, choose a particulr section. When $y = 4$, for example, we get $z = -0.5x + 12 + c$. In Figure 2.7b, we can see that the $y = 4$ contour passes through $x = -4$ and $z = 0$. Substitute to get $c = -14$. Therefore, the equation of this surface is $z = -0.5x + 3y - 14$. This is the equation of a plane intersecting the yz-plane in the line $z = -0.5x - 14$, the yz-plane in the line $z = 3y-14$, and the xy-plane in the line $-0.5x + 3y = 14$.

Problems

1. What are the x-sections and y-sections of the plane $z = x + 2y$?

2. Describe the x-sections and y-sections of the surface $x^2 + y^2 + z^2 = 25$.

3. **a.** Sketch the x-sections of the surface $x^2 + y^2 = z^2$ for $x = -2, -1, 0, 1,$ and 2.

   b. Describe the y-sections of the surface.

   c. What does the surface look like?

4. **a.** Sketch the x-sections of the surface $x^2 + y^2 = z$ for $x = -2, -1, 0, 1,$ and 2.

   b. Describe the y-sections of the surface.

   c. What does the surface look like?

5. The z-sections of the surfaces in problems **3.** and **4.** are circles centered at the origin. How do the z-sections in problem **3.** differ from the z-sections in problem **4.**?

6. Sketch a typical x-section and a typical y-section of the surface $(x + y)^2 = z$. What is the intersection of this surface with the plane $z = 4$? $z = 1$? Sketch the surface.

7. Describe the x-sections and y-sections of the surface $x^2 - y + z^2 = 0$.

8. Find a multivariable function whose x-sections are parabolas and whose y-sections are straight lines.

9. Find a multivariable function whose x-sections are lines and whose y-sections are hyperbolas.

10. Find a multivariable function whose x-sections and y-sections are both hyperbolas. What are the z-sections of your function?

11. Find a multivariable function whose x-sections are lines and whose y-sections are exponential decay functions. What are the z-sections of your function?

12. Sketch, by hand, x-sections of the function $z = \sqrt{x + y}$ for $x = -6, -4, -2, 0, 2, 4,$ and 6.

13. The general linear function in the plane is $ax + by = c$. The general linear function in space is $ax + by + cz = d$. The x-sections and y-sections of a particular linear function are shown in Figures 2.8a and 2.8b, respectively. Find the values of a, b, c, and d.

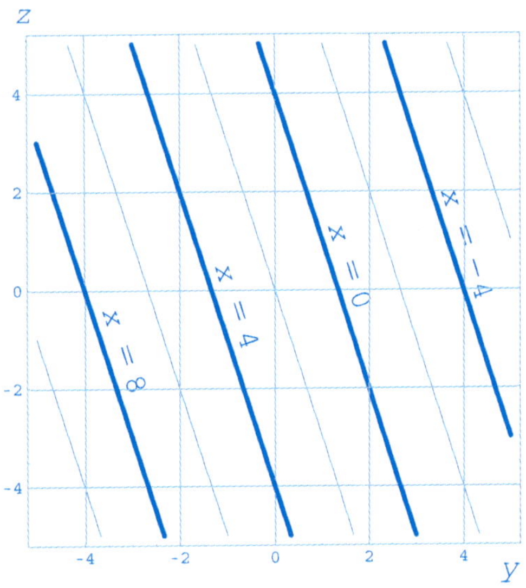

Figure 2.8a

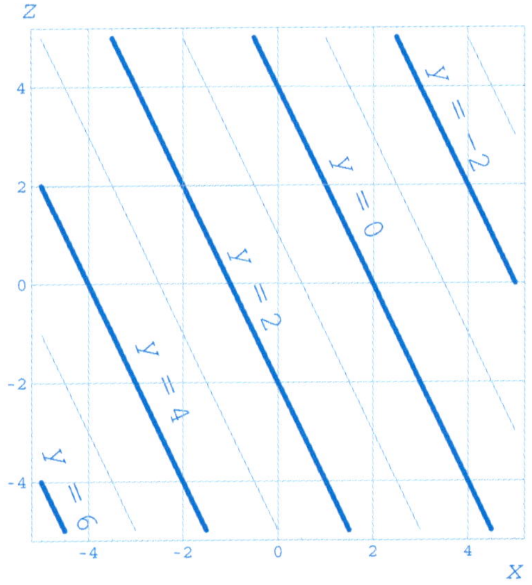

Figure 2.8b

14. The x-sections and y-sections of a particular linear function are shown in Figures 2.9a and 2.9b, respectively. Find the values of a, b, c, and d.

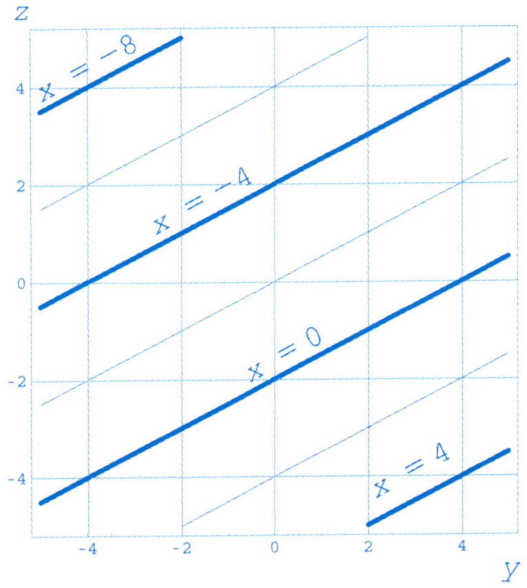

Figure 2.9a

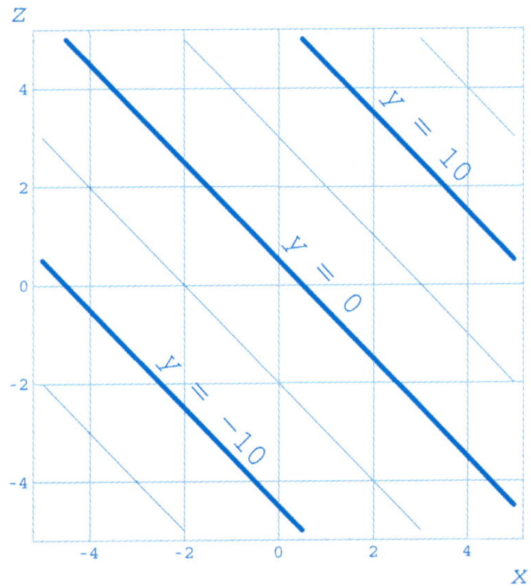

Figure 2.9b

15. Some multivariable functions borrow their names from similar single-variable functions of calculus. For example, $y = mx + b$ is a linear function because the highest power of the variables is 1 and so the analogous two-variable linear function is $z = ax + by + c$. The perimeter of a rectangle $P = 2L + 2W$, for example, is a linear multivariable function. The surface is shown in Figure 2.10. What is the domain of the perimeter function? What meaning can you attach to the L-sections, W-sections, and P-sections?

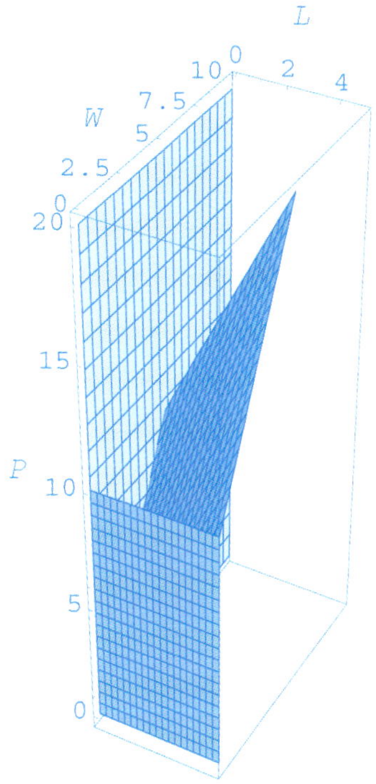

Figure 2.10

The next two problems are about quadratic multivariable functions. The general quadratic function in a single variable is $y = ax^2 + bx + c$. The corresponding two-variable quadratic function is $z = ax^2 + by^2 + cxy + dx + ey + f$.

16. Notice that the general form includes an xy term. That means that $z = xy$ is considered a quadratic surface even though neither variable is squared. This is reasonable because this function could represent, for example, the area of a rectangle. $A(L, W) = LW$ gives the area of a rectangle and, while L and W are both linear, the area is measured in square units. The surface $A = LW$ is shown in Figure 2.11.

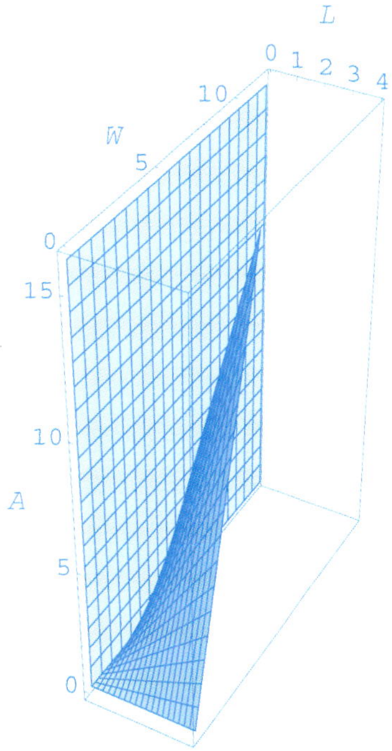

Figure 2.11

a. What is the domain and range of the area function?

b. What are the L-sections and W-sections of this area function? How can we interpret them in geometric terms?

c. What are the A-sections of this function? Interpret them in geometric terms.

17. Figures 2.12a and 2.12b show x-sections and y-sections of a particular quadratic surface. The highest y-section is $z = x^2 - 16$ and the lowest x-section is $z = -0.5y^2 - 16$. What are the values of a, b, c, d, e, and f?

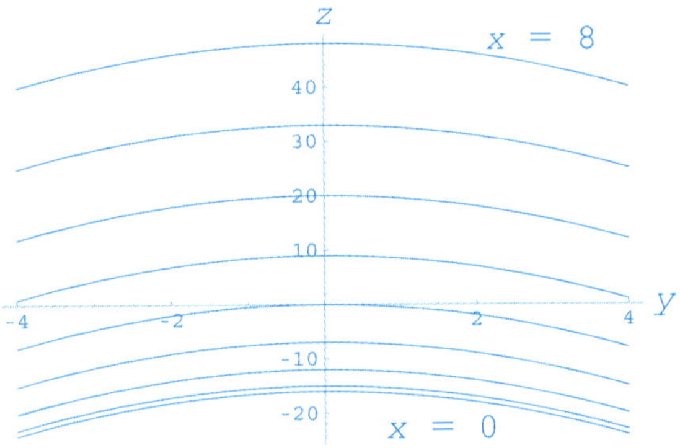

Figure 2.12a

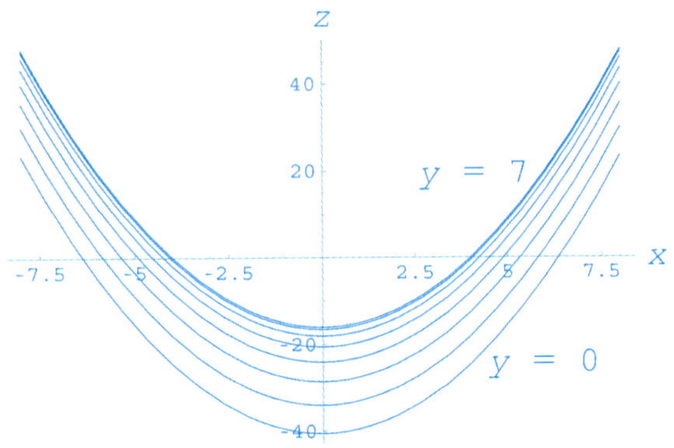

Figure 2.12b

18. You may be familiar with the greatest integer function $y = [x]$. This is sometimes called a step function or a staircase function and many calculators name it *int*(x). Whatever the name, it represents a function that outputs the greatest integer that is not larger than x. For example, $[7.3] = 7$, $[3] = 3$, and $[-\pi] = -4$. Figure 2.13 shows a graph of $y = [x]$ on $[-3, 3]$; it is composed of horizontal line segments. Notice that the endpoints of each segment are marked to show the jump from one level to the next.

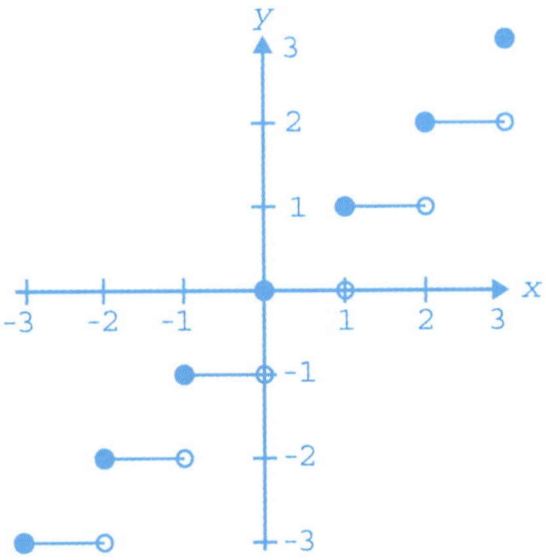

Figures 2.13

It is an interesting challenge to lift this function out of the plane and examine it in space.

a. Describe the surface $z = [x]$, which is composed of many infinitely long steps. What becomes of the open and closed endpoints of Figure 2.13 as we move into three dimensions? Try to sketch this discontinuous surface for $-3 \le x \le 3$.

b. Describe the surface $z = [x + y]$. This is composed of square platforms. Describe a typical square. Try to sketch this surface for $0 \le x \le < 3$ and $0 \le y \le 3$.

c. Describe the surface $z = [x] + [y]$. This is also composed of many infinitely long steps. Describe a typical step. Try to sketch this discontinuous surface for $-3 \le x \le 3$, $-3 \le y \le 3$.

CHAPTER 3

Visualizing Surfaces with Contour Diagrams

Functions of one variable, $y = f(x)$, graph as curves in the plane while functions of two variables, $z = g(x, y)$, graph as surfaces in space. In the last chapter we saw how the x- and y-sections of a three-dimensional surface could help us to visualize the surface. In this chapter we focus on z-sections, which are called *level curves or contours*, to help create an image of a three-dimensional surface. A collection of contour graphs is called a *contour diagram*.

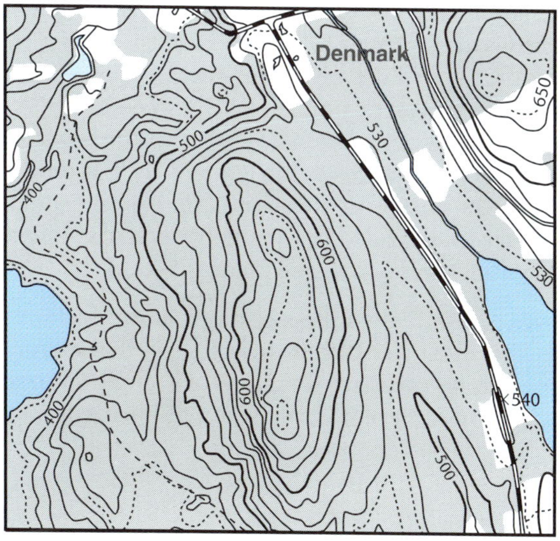

Figure 3.1

Figure 3.1 is a portion of a topographical map of a small area of Maine. Such survey maps help those who know how to read them understand the contours of the land and therefore are called *contour maps or topographical maps*. The most obvious feature of this map is its many strange loops and whorls. These curves, or *contours*, represent elevation, that is, the height of the land above sea level. Note that the darker contours are numbered. All points on the 600-foot contour, for example, are 600 feet above sea level. Between each pair of dark contours are four lighter ones, which represent intermediate elevations. On this map, each contour represents a 20-foot change in elevation.

One way to understand the curves on a contour map is to imagine a great flood that covers all the land. As the water recedes, the highest points appear as islands in this imaginary sea. As the water level drops, the islands grow larger until they merge with each other. The contours mark the change in water level at 20-foot intervals.

Reading the Contours

The lowest land shown in figure 3.1 is on the left edge of the map. As we move to the right, the elevation increases. Contours that are bunched up indicate rapid elevation changes, while contours that are widely separated indicate more gradual elevation changes. A person in the region indicated by the bottom center portion of the map who is walking due north 5,000 feet would rise about 140 feet. On the other hand, a person in the region indicated by the middle of the left edge of the map who is walking east would rise about 60 feet over the same distance. Mathematically, these changes in elevation correspond to slope. The slope heading north from the bottom is about $\frac{140 \text{ feet}}{5000 \text{ feet}} \approx 0.028$, while the slope heading east from the left edge is about $\frac{60 \text{ feet}}{5000 \text{ feet}} = 0.012$.

Contour Diagrams

Figure 3.2 shows a contour diagram for the function $z = g(x, y) = x^2 - x + y^2 + xy^2$. The diagram is very similar to a topographical map such as the one we just examined. If we think of the z-values of the function as elevations above or below the xy-plane, then the curves represent points at the same elevation. These curves are called *level curves of g*, or simply contours. The loop near the origin suggests a spike in the surface. The separation of the curves to the left tells us that the surface is less steep there, while the density of the curves in the right hand corners suggests that the surface is steeper there. The curves appear to be symmetrical across the x-axis, which suggests that the surface is symmetrical across the plane $y = 0$.

Level curves are similar to the x-sections and y-sections of a surface we examined in Chapter 2. The most significant difference is that x- and y-sections of a multivariable function are always ordinary functions of a single variable, but z-contours typically are not. They usually represent implicit equations involving both x and y.

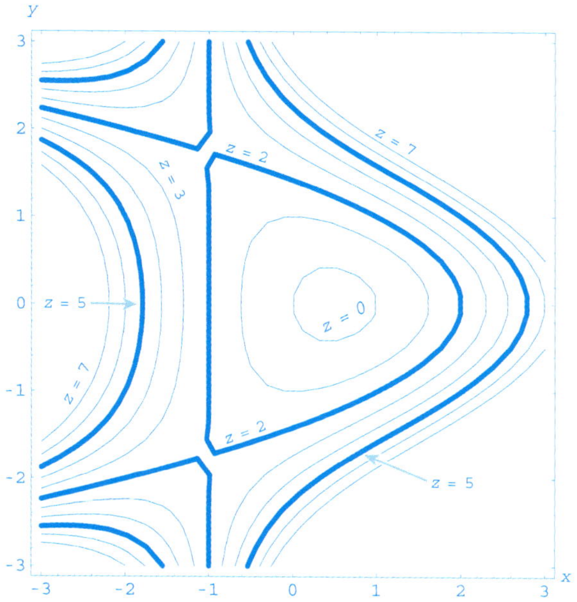

Figure 3.2

Example 1 Draw level curves for the plane $z = 2x + y - 1$ in the window $[-3, 3] \times [-6, 6]$.

Solution Each particular value of z produces a linear function $y = -2x + (z + 1)$ with slope -2 and y-intercept $z + 1$. Figure 3.3 shows these levels.

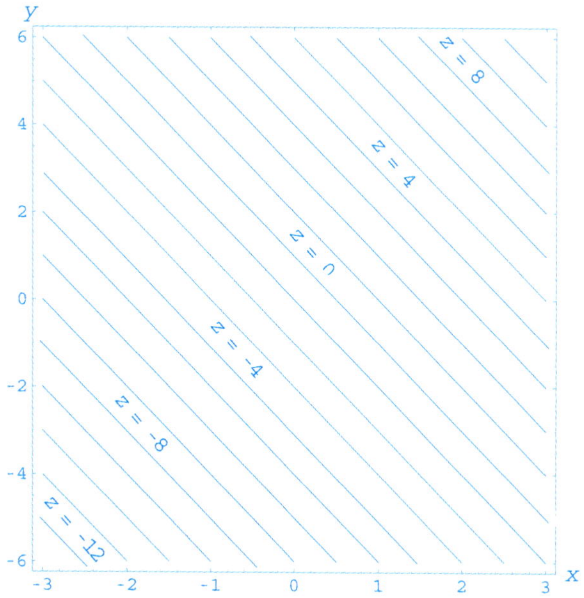

Figure 3.3

There is much information contained in this figure. For example, the point $(-1, -6)$ on the level curve $z = -9$ tells us that the point $(-1, -6, -9)$ lies on the surface $z = 2x + y - 1$. The points $(-3, -2)$ and $(-3, 4)$ on their level curves tell us that the points $(-3, -2, -9)$ and $(-3, 4, -3)$ lie on the surface. We can conclude that as we increase the y-coordinate by 6, we increase z-coordinate by 6 as well. That is, as we hold our x-coordinate constant and move along the surface increasing our y-coordinate, we move higher on the surface, increasing our z-coordinate.

Notice that the contours are equally spaced. This tells us that equal changes in x produce equal changes in z and, in the same way, equal changes in y produce equal changes in z. We will examine this type of information in much greater detail in the next chapter.

Example 2 Describe the level curves of the plane $x - 3y + 3z = 15$.

Solution Treat z as constant and solve for y. We see that the equation of a typical level curve is $y = \frac{1}{3}x + (z - 5)$. A set of level curves is shown in Figure 3.4.

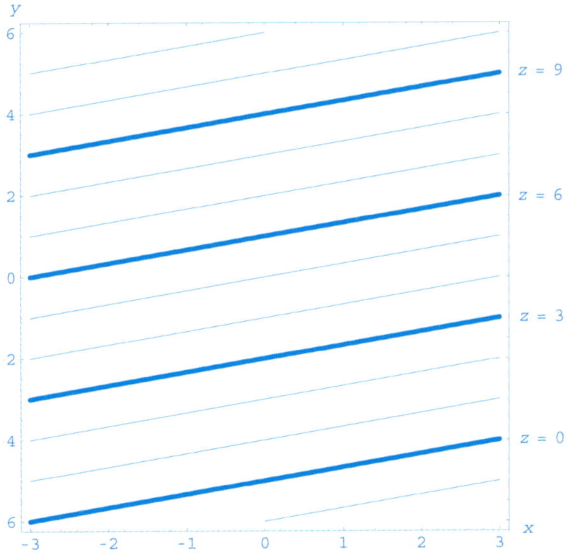

Figure 3.4

Note that the contours are parallel and evenly spaced. The contours are parallel because they each have slope $\frac{1}{3}$. They are evenly spaced because each change in z produces the same change in the y-intercept. For example, each time z changes by 1, the level curve shifts 1 unit in the positive y-direction.

Planes are not the only surfaces that can have linear contours, as the next example illustrates.

Example 3 Describe the level curves of $z^3 = 2x + 3y$.

Solution For any particular value of z, the associated level curve is given by the formula $2x + 3y = z^3$. This formula is linear in x and y. If we write these level lines as $y = -\left(\frac{2}{3}\right)x + \frac{z^3}{3}$, we can see that the contours are parallel lines with slope $-\frac{2}{3}$. In addition, since the y-intercept of these levels is a cubic, $\frac{z^3}{3}$, it follows that as z increases the intercept increases at an ever-increasing rate. This can be seen in Figure 3.5 where the gaps between contours grow wider as z gets larger. While these contours are linear, they are *not* equally spaced.

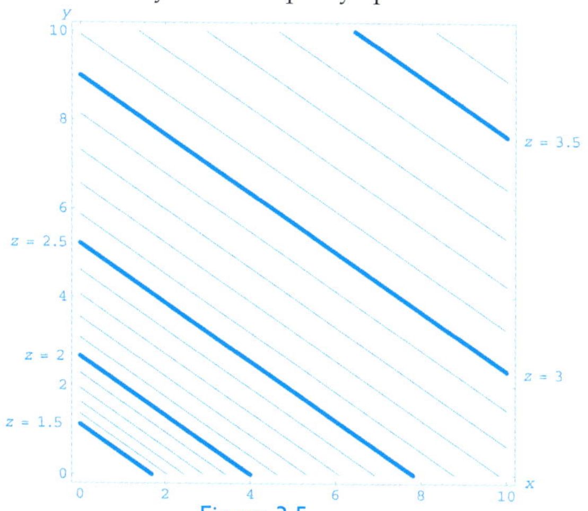

Figure 3.5

Lines are not the only contours that can be equally or unequally spaced. The next example shows functions whose level curves are concentric circles.

Example 4 Draw the level curves of the surface $z^2 = x^2 + y^2$ for $z = 1, 2, 3, 4, \ldots$ and explain why they are equally spaced.

Solution A typical contour is $x^2 + y^2 = z^2$, a circle centered at the origin with radius z. These level curves are shown in Figure 3.6. Since each unit increase in z increases the radius by 1, the contours are equally spaced concentric circles. By the way, the $z = 0$ contour (not shown) is a single dot at the origin.

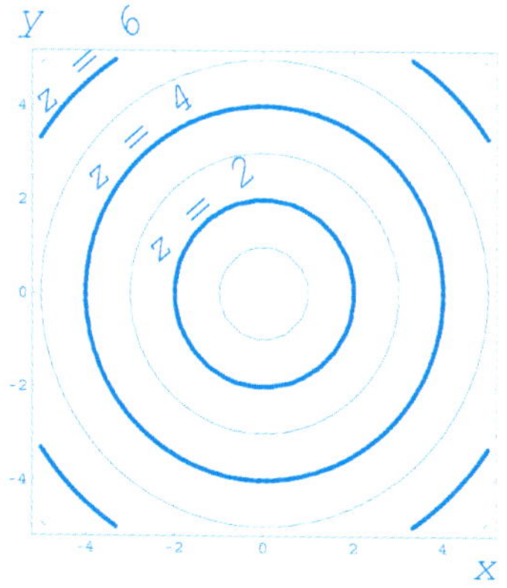

Figure 3.6

The next example illustrates a function whose z-sections are also concentric circles, but not equally spaced.

Example 5 Describe the level curves of the surface $z = x^2 + y^2 - 4$.

Solution The contours are of the form $x^2 + y^2 = z + 4$, representing circles with radius $\sqrt{z + 4}$ centered at the origin. Figure 3.7 shows that the level curves become denser as we move away from the origin and sparser as we move toward the origin. That is, as z increases, the radii of the circle increase *but not at a constant rate.* The surface rises up more and more steeply as we move away from the center of these contours.

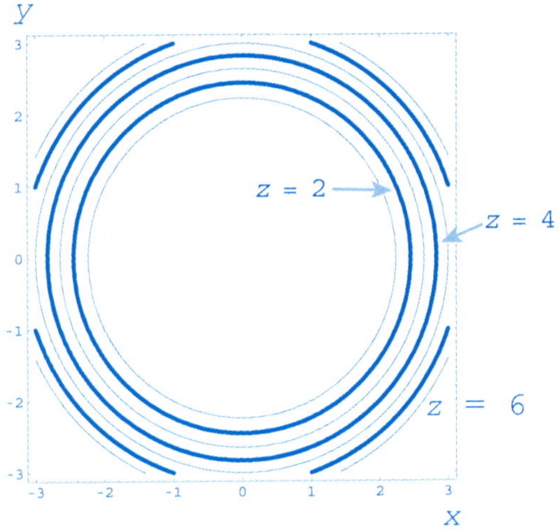

Figure 3.7

Level Curves and Implicit Functions

You have seen level curves before, but in a very different context. For example, the $z = 6$ level curve of the multivariable function $f(x, y) = (y - x - 4)(y + 2)(x - 3)$ is the implicit equation $6 = (y - x - 4)(y + 2)(x - 3)$ shown in Figure 3.8.

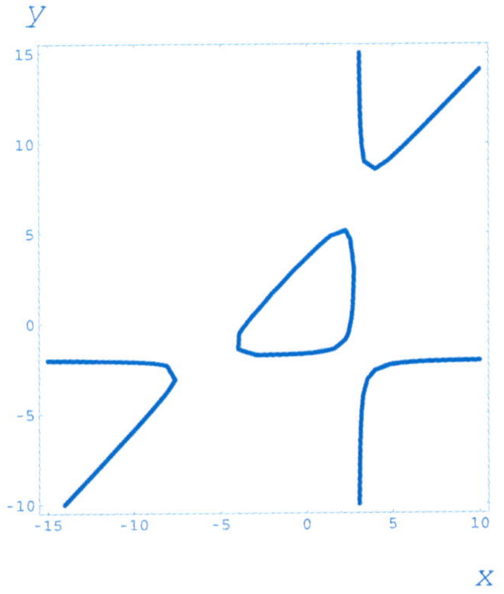

Figure 3.8

Perhaps, when you were learning implicit differentiation, you wondered at the strange shapes of the graphs of implicit equations, which were so much more complex and elaborate than the simpler functions you had been used to seeing. Now we have a new context for understanding this complexity. Figure 3.9 shows a part of the surface of $z = (y - x - 4)(y + 2)(x - 3)$ and the plane $z = 6$. Notice how the oval in Figure 3.8 can now be seen as a level curve of a central spike.

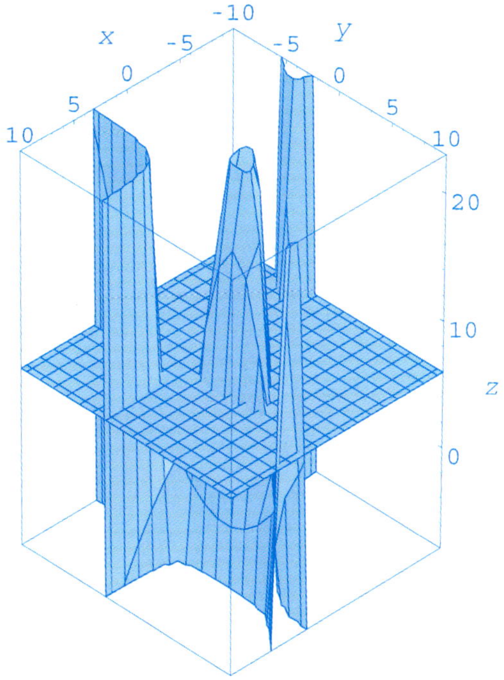

Figure 3.9

The three-dimensional view shows us not only that the central oval is an intersection of $z = 6$ with a spike in the surface, but also that as we continue to increase z, the oval will shrink and eventually vanish when the z-level is greater than the height of the spike. As z decreases, the oval reappears and eventually merges with the three hyperbola-like branches of this surface.

Problems

1. Which of the graphs in Figures 3.10a and 3.10b could be the level curves of $z = ax + by$ if $a > 0$ and $b < 0$?

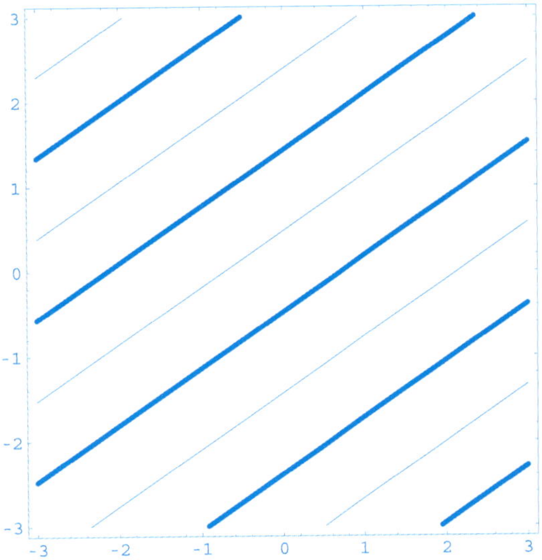

Figure 3.10a

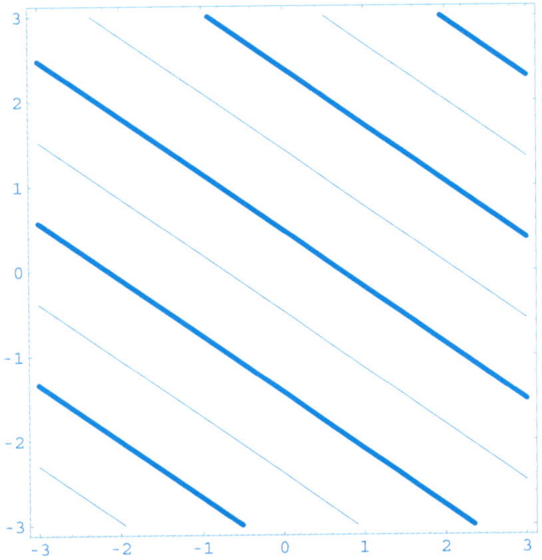

Figure 3.10b

2. Match the equation to its contour diagram.

   i. $z^2 = 2x^2 + y^2$ ii. $z^2 = x^2 + y$

   iii. $-x + \frac{y}{2} - 4 = z$ iv. $\frac{x}{2} - y + 4 = z$

   v. $z^2 = x^2 + y^2$ vi. $z^2 = x^2 + 2y^2$

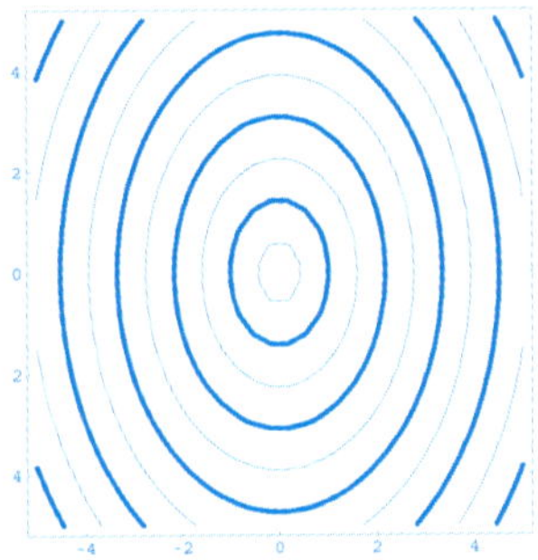

Figure 3.11a

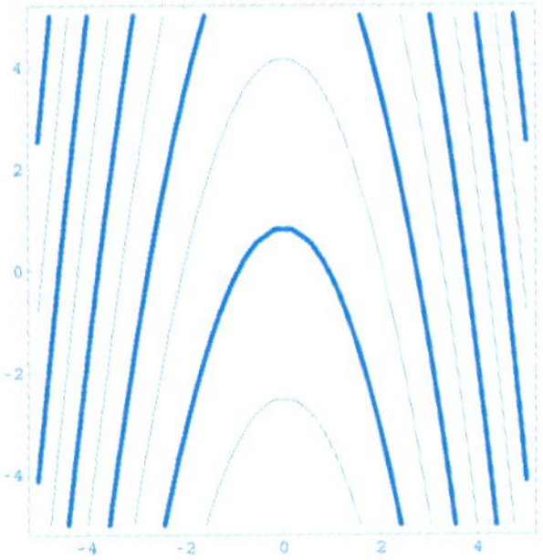

Figure 3.11b

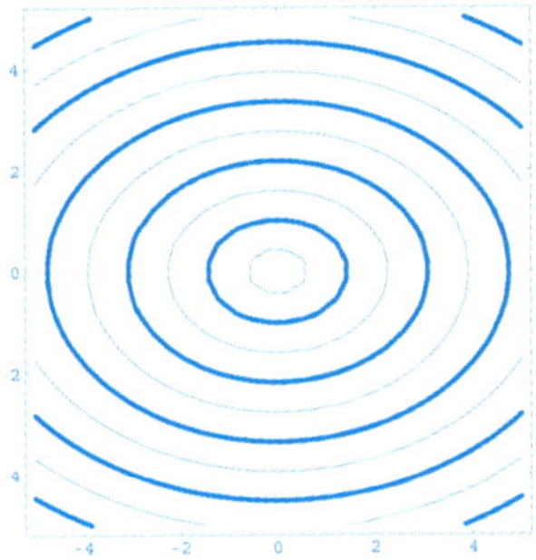

Figure 3.11c

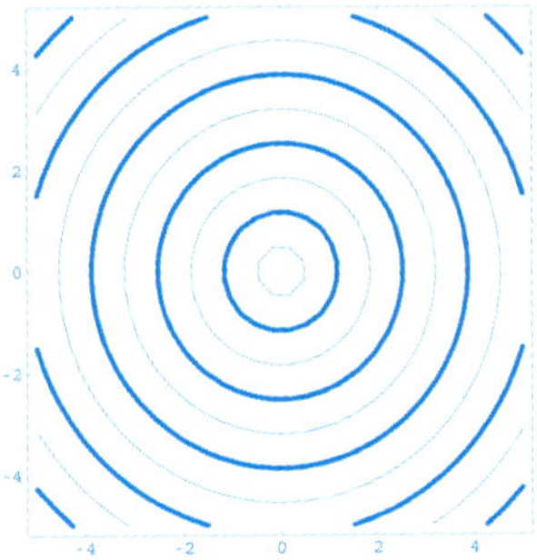

Figure 3.11d

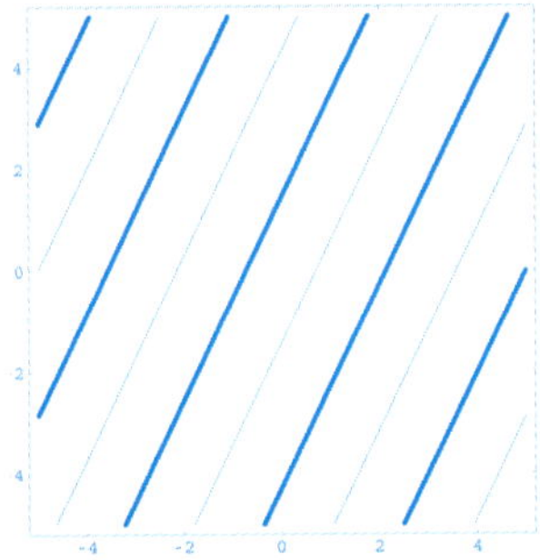

Figure 3.11e

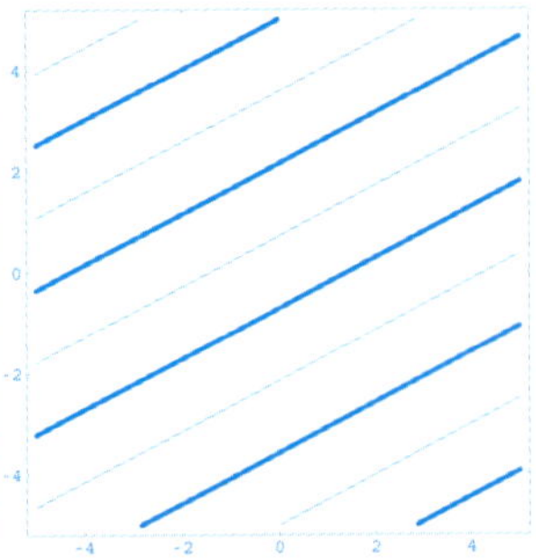

Figure 3.11f

3. Match the equation to its contour diagram.

   i. $x + y = z$

   ii. $(x + y)^2 = z$

   iii. $(x + y)^2 = z^3$

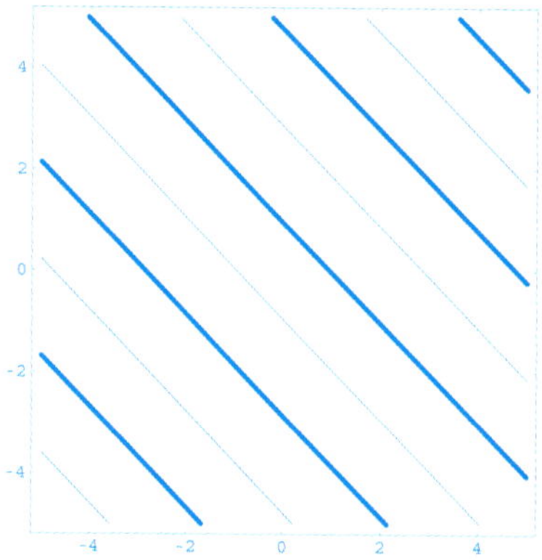

Figure 3.12a

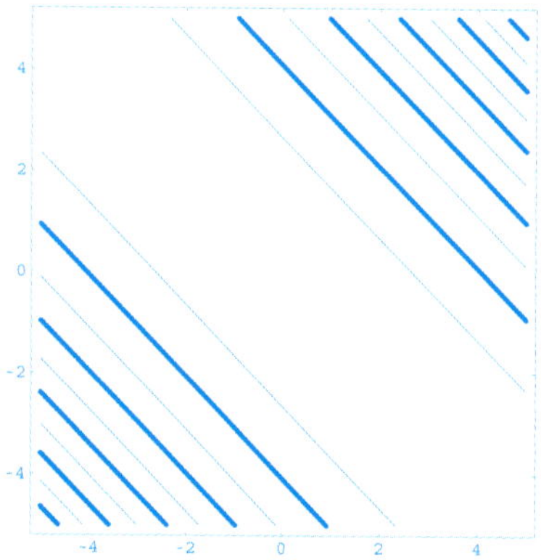

Figure 3.12b

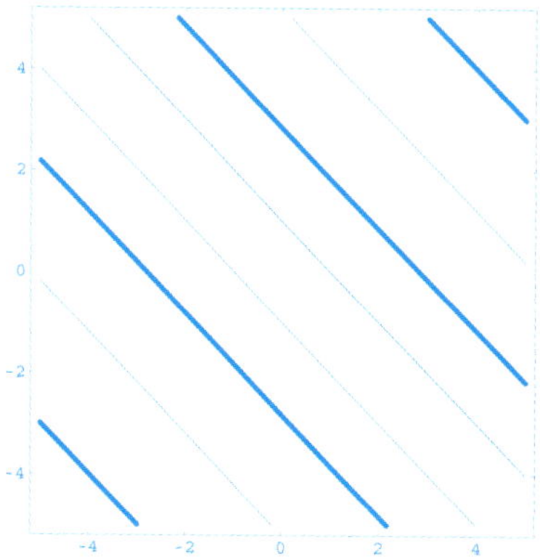

Figure 3.12c

4. Describe the level curves for $z = x^3 - y$ for $z = 0, 1, 2, \ldots$ What are their shapes? Are they evenly spaced? If not, do they grow denser or sparser as z increases?

5. **a.** Find an equation of a surface whose level curves are equally spaced sinusoids for $z = 0, 1, 2, \ldots$

   b. Find an equation of a surface whose level curves are sinusoids that grow denser as z increases.

6. The single-variable exponential decay function $y = e^{-x}$ is asymptotic to the x-axis. When we treat this function as a multivariable function, its graph is a surface in space. Describe the surface. Is there still an asymptote? If so, what is it?

7. Figure 3.13 shows the surface $z = \dfrac{2 + x + y}{x + y}$, which appears to be asymptotic to two planes. What are they?

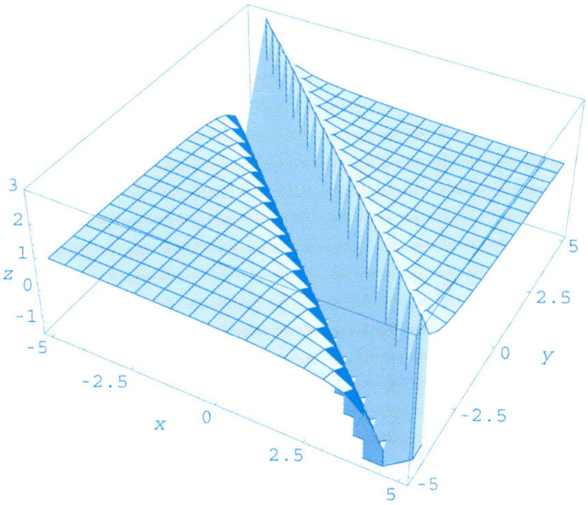

Figure 3.13

8. Figure 3.14 shows the surface $z = \frac{1}{x} + \frac{1}{y}$, which appears to have four branches, each asymptotic to two planes. What are the asymptotic planes?

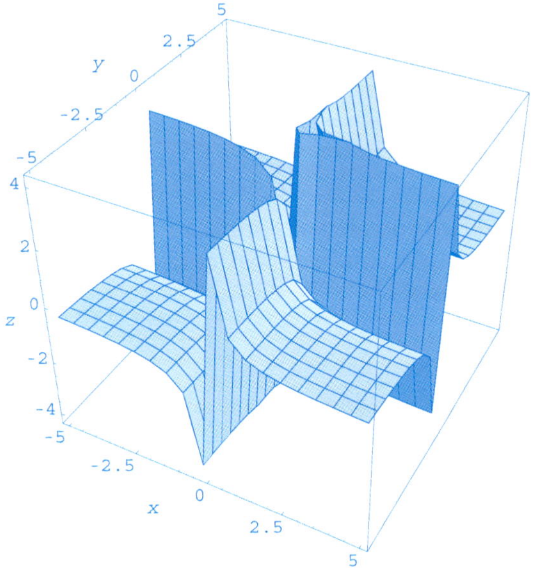

Figure 3.14

9. The function $z = [x] - [y]$ does not have level curves, but rather level regions. (Recall that $y = [x]$ is the greatest integer function.) Sketch the level surfaces when $z = -1, 0, 1$, and 2 and describe the overall surface.

CHAPTER 4

Differentiation in Space: Partial Derivatives

In this chapter we will investigate what happens to the concept of a derivative when we leave the plane and move into space. The derivative of a single-variable function represents the slope of a tangent line to the curve at a point; it is the rate at which the y-values change as we change the input values x. We can visualize the derivative by imagining that we are riding along a two-dimensional curve $y = f(x)$. As we move, we increase or decrease our x-coordinate. By moving to the left or to the right, we create a corresponding change in our y-coordinate and so we rise and fall with the curve. In space, however, things are very different.

Imagine instead that we are on a hillside. We can move in the east-west direction or in the north-south direction or in any other direction we choose, and each movement will change our elevation. This is the same observation we made in Chapter 3 when we examined the topographic map of Maine.

If, instead of a hillside, we are on the surface $z = g(x, y)$, then we could move in the x-direction, which would change our x-coordinate but not our y-coordinate, or we could move in the y-direction, which would change our y-coordinate but not our x-coordinate. We could also move in a direction that would change both our x- and our y-coordinates. In every case, as our input coordinates change, our z-coordinate also changes and so we move up or down on the surface.

We can be more precise by introducing Δx and Δy to represent the amounts by which we change the input variables. In two dimensions, we have only two possibilities. If we move to the right, then we increase x and Δx is positive; if we move to the left, then we decrease x and Δx is negative. With three dimensions, however, not only can we increase or decrease both x and y, but we can also change them in unequal amounts.

Therefore, when we choose a direction on the surface in which to move, we are choosing a Δx and a Δy. Changing just one of these will move us in the direction of the x-axis or the y-axis. Changing both Δx and Δy, however, will move us in some other direction. This direction is usually specified by the vector $\Delta x\vec{i} + \Delta y\vec{j}$, however, in this book we will not work with vectors. It is enough to realize that changing both x and y will correspond to moving in some direction other than along the x- or y-axis, and these changes will produce some change in height Δz.

In single-variable calculus, the change in y produced by a change in x approximates the derivative of a function. In multivariable calculus, the rate at which z changes will, in general, depend upon the *direction* chosen and the resulting derivative is called the *directional derivative*. In this book, we will restrict our attention to the rate at which z changes when we change only x or only y but not both together. This leads to two new types of space derivatives called *partial derivatives*. More will be said about them later. For now, we will see how level curves can help us think about these rates of change.

Example 1 Figure 4.1 shows level curves for the plane $z = 3x + y$. Use these curves to estimate the rate of change of z with respect to x and the rate of change of z with respect to y at the point (1, 1, 4).

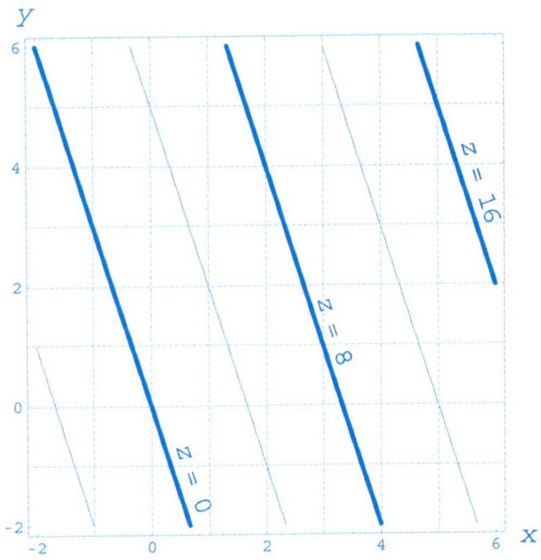

Figure 4.1

Solution The point (1, 1) lies on the $z = 4$ contour. If we move horizontally to the next contour, we do not fall exactly on a grid point and would have to approximate the coordinates of the point, which are somewhere around (1.7, 1). That is, when $\Delta x \approx 0.7$, $\Delta z = 2$ and so the rate of change of z with respect to x at (1, 1) is approximately $\frac{20}{7}$. We can get a more accurate result, in this example, in two different ways. First of all, we can notice, for example, that by moving 2 units to the right we fall exactly on the $z = 10$ contour, so that $\frac{\Delta z}{\Delta x} = \frac{6}{2} = 3$. We can also compute this value exactly from the equation. Since $z = 3x + y$, moving to the point (2, 1) gives $z = 3(2)+1 = 7$ and, again, the rate of change of z with respect to x is 3.

If, instead, we change y by 2 and move from (1, 1) to (1, 3), we move from $z = 4$ to $z = 6$. Therefore, $\frac{\Delta z}{\Delta y} = 1$ and the rate of change of z with respect to y at (1, 1) is 1. Since $z = 3x + y$ represents a plane and the level curves for a plane are equally spaced, the fact is that these slopes will be the same at any point, not just at (1, 1, 4).

Notice that the slopes we found are the coefficients of x and y. This is not a coincidence. We shall see shortly that this result is analogous to the fact that, in the familiar slope-intercept form of a line in the plane, $y = mx + b$, the coefficient of x is the slope of the line.

The next example shows that the slopes on a more complex surface are not the same at each point but will, in general, depend upon the particular point we choose.

Example 2 Figure 4.2 shows level curves for a particular function. Approximate the rate of change of the surface in the x and in the y directions at $(1, -1)$.

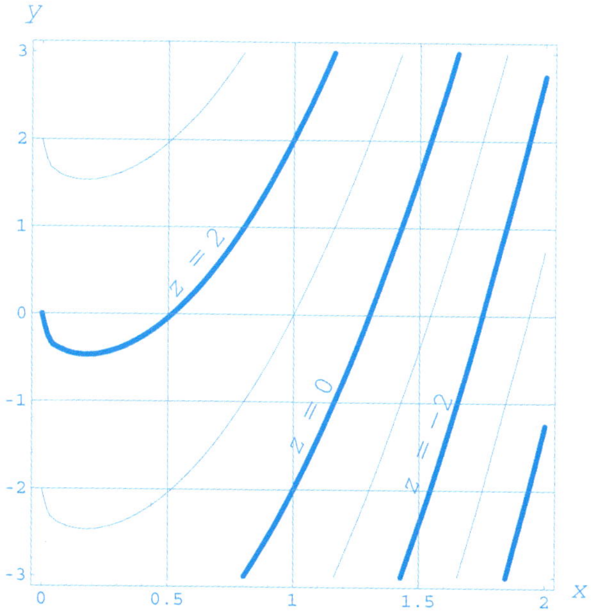

Figure 4.2

Solution This function is quite a bit more complicated than the function in the previous example. How can we get the most accurate estimate in this situation? If this were a single-variable function, what would be the best strategy to use? For example, suppose we had the function whose values are given in Table 4.1 and we wanted to estimate the derivative at $x = 2$.

x	1	2	3
y	2.5	3	3.2

Table 4.1

There are three ways that we could do this. First, we could compute the average rate of change of the function from $x = 1$ to $x = 2$ and estimate $y'(2) = 0.5$. This estimates "the slope from the left." Alternatively, we could compute the average rate of change of y from $x = 2$ to $x = 3$, the "slope from the right," and estimate $y'(2) = 0.2$. The best estimate, however, would be obtained by using the average rate of change from $x = 1$ to $x = 3$, and therefore estimate 0.35. Note that this method won't work if the function fails to have a derivative somewhere in the interval $1 \leq x \leq 3$, for example, if it has a vertical asymptote in this interval. But, if the graph is differentiable on this interval, then this third method gives the best estimate of the slope at $x = 2$.

Similar reasoning works with multivariable functions. To estimate the rate of change of z with respect to x at $(1, -1)$, we move horizontally through $(1, -1)$ to each of the nearest contours. In this case, the level curve to the left is also $z = 1$. This curve intersects $y = -1$ at, approximately, $x = 0.4$. To the right, the line $y = -1$ intersects the level curve $z = 0$ at approximately $x = 1.6$. This gives us the points $(0.4, -1)$ and $(1.6, -1)$, which correspond to the points $(0.4, -1, 1)$ and $(1.6, -1, 0)$ on the surface. The average rate of change of z with respect to x between these two points is

$$\frac{\Delta z}{\Delta x} = \frac{0 - 1}{1.6 - 0.4} = -\frac{5}{6}.$$

Similarly, to estimate the average rate of change of z with respect to y, move along the vertical line $x = 1$ between the surrounding level curves to the points $(1, -2)$ and $(1, 0)$, which correspond to the points $(1, -2, 0)$ and $(1, 0, 2)$ on the surface. This gives an average rate of change of z with respect to y of $\frac{\Delta z}{\Delta y} = \frac{2 - 0}{0 + 2} = 1$. Therefore, if we move in the direction of increasing x, we move lower on the surface, while if we move in the direction of increasing y, we move higher on the surface.

Example 3 Table 4.2 gives the values of some multivariable function $z = g(x, y)$ in the domain $[0, 1] \times [0, 1]$. Use the table to approximate the rate of change of z in both the x and y-directions at $(0.6, 0.4)$ and interpret the results

x \ y	0.2	0.4	0.6	0.8	1
0.2	−.04	0.12	0.36	0.68	1.08
0.4	−0.36	−0.16	0.12	0.48	0.92
0.6	−0.92	−0.68	−0.36	0.04	0.52
0.8	−1.72	−1.44	−1.08	−0.64	−0.12
1	−2.76	−2.44	−2.04	−1.56	−1

Table 4.2

Solution Consider just the second row of this table. By holding y constant, we are restricting our attention to a y-section of this surface and the values in this row are the values of a single-variable function of x. We approximate the rate of change of this function at $x = 0.6$ by using the (x, z) values $(0.8, 0.48)$ and $(0.4, -.16)$. This gives an approximation for the rate of change of z with respect to x at $(0.6, 0.4)$ as $\frac{\Delta z}{\Delta x} \approx \frac{-.16 - .48}{0.4 - 0.8} = 1.6$.

Similarly, we approximate the rate of change at $y = 0.4$ while keeping x constant by using the (y, z) points $(0.2, .36)$ and $(0.6, -.36)$. This gives $\frac{\Delta z}{\Delta x} \approx \frac{.36 + .36}{0.2 - 0.6} = -1.8$. We can interpret these numbers by imagining a bug sitting on the surface at $(0.6, 0.4, 0.12)$ above the point $(0.6, 0.4)$ in the xy-plane. As the bug moves towards the point above $(0.6, 0.6)$, while keeping $x = 0.6$, it will move down the surface at approximately 1.8 z-units for each unit change in y. If it moves toward the point above $(0.8, 0.4)$, holding y constant, then it will move up the surface at 1.6 z-units for each unit change in x.

Derivatives with Directions: Partial Derivatives

In the previous examples, we repeatedly spoke of "moving in the x-direction," or of "moving in the y-direction." If we move in the x-direction then we are holding y constant, so moving in the x-direction means moving within some y-section of the surface. But moving in a section means moving along a two-dimensional curve in a plane, and so we might expect the rate of change to be the ordinary derivative of that particular function. This is exactly the tool we will use to measure rates of change. In general, the value of this derivative will depend upon the particular x-section or y-section that we choose. In order to distinguish this rate of change from the ordinary derivative in two dimensions, we introduce the new symbol $\frac{\partial z}{\partial x}$ for the rate of change of z with respect to x. Here x is changing and y is being held constant. Since this tells only part of the story of how the surface is changing, it is called the *partial derivative of z with respect to x*. Similarly, we write $\frac{\partial z}{\partial y}$ for the *partial derivative of z with respect to y* while holding x constant.

Example 4 Compute $\frac{\partial z}{\partial x}$ and $\frac{\partial z}{\partial y}$ for the function $z = y - x^2 + 3x^{\frac{1}{3}}$ and evaluate these derivatives at $(x, y) = (1, -1)$ and $(3, 1)$.

Solution This is the same function we examined in Example 2, but in Example 2 we only approximated the rates of change. Now we will see how to find the exact values by differentiation. To find the partial derivative of z with respect to x, treat y as if it were a constant and differentiate. Since the derivative of a constant is 0, we get $\frac{\partial z}{\partial x} = -2x + \frac{1}{x^{\frac{2}{3}}}$. Therefore, $\left.\frac{\partial z}{\partial x}\right|_{(1,-1)} = -1$ and $\left.\frac{\partial z}{\partial x}\right|_{(3,1)} = -2(3) + \frac{1}{3^{\frac{2}{3}}} \approx -5.52$.

In the same way, if we hold x constant, we get $\frac{\partial z}{\partial y} = 1$ for all values of x.

Because it is awkward to indicate inputs to the derivative when using this notation, there is an alternate functional notation for partial derivatives. The partial derivative of g with respect to x is also written $g_x(x, y)$ and the partial derivative of g with respect to y is written $g_y(x, y)$. Using this notation we would write the results of Example 4 as $g_x(1, -1) = -1$ and $g_x(3, 1) \approx -5.52$.

While this new notation may seem a little strange, the process is not. All the rules for differentiation that you know still apply. Just remember to treat x as a constant when finding g_y, since we are in an x-section of the surface, and similarly to treat y as constant when finding g_x.

Example 5 Compute $f_x(-2,0)$ and $f_y(-2, 0)$ for $f(x, y) = x^2\tan\left(\frac{\pi e^y}{4}\right) - 2y + 5$.

Solution Hold y constant to compute $f_x(x,y) = 2x\tan\left(\frac{\pi e^y}{4}\right)$, so that $f_x(-2,0) = -4$.

Hold x constant to compute $f_y(x, y) = \dfrac{x^2\sec^2\left(\frac{\pi e^y}{4}\right)\pi e^y}{} - 2,$

so that $f_y(-2, 0) = \pi\sec^2\left(\frac{\pi}{4}\right) - 2 = 2\pi - 2.$

The next example shows that we must sometimes use the product and quotient rules when computing partial derivatives.

Example 6 Compute the partial derivatives of $z = \frac{x^3e^{xy}}{y+2}$.

Solution Hold y constant and use the product rule to compute $\dfrac{\partial z}{\partial x} = \dfrac{x^3ye^{xy} + 3x^2e^{xy}}{y+2}$.

Hold x constant and use the quotient rule to compute $\dfrac{\partial z}{\partial y} = \dfrac{(y+2)x^4e^{xy} - x^3e^{xy}}{(y+2)^2}$.

Equations of Planes

In Example 1, we saw that the slope of $z = 3x + y$ in the x-direction was 3 and the slope in the y-direction was 1. With the aid of partial derivatives, this result is now obvious since $\frac{\partial z}{\partial x} = 3$ and $\frac{\partial z}{\partial y} = 1$. Now let's generalize this result. The general linear function in space is $ax + by + cz = d$, which we can rewrite as $z = -\frac{a}{c}x - \frac{b}{c}y + \frac{d}{c}$ and $\frac{\partial z}{\partial x} = -\frac{a}{c}$ and $\frac{\partial z}{\partial y} = -\frac{b}{c}$. Furthermore, if $x = y = 0$, then $z = \frac{d}{c}$. That is, the general equation of a plane can be written as $z = m_x x + m_y y + b_z$ where m_x is the slope in the x-direction, m_y is the slope in the y-direction, and b_z is the z-intercept. This is exactly analogous to the slope-intercept form of a two-dimensional line $y = mx + b$. We might call this the *slope-intercept form of a plane*. Of course, there are actually two slopes this time, one for each partial derivative.

The other important form of a line is the point-slope form $y - y_0 = m(x - x_0)$, which gives the equation of a line passing through (x_0, y_0) with slope m. Similarly, the *point-slope form of a plane* is $z - z_0 = m_x(x - x_0) + m_y(y - y_0)$ where m_x and m_y are the slopes and (x_0, y_0, z_0) is the point. We can easily derive this form by taking the equation of a plane passing through the origin with slopes m_x and m_y, which has the equation $z = m_x x + m_y y$ (since the z-intercept is zero), and shifting it x_0 units in the x-direction, y_0 units in the y-direction, and z_0 units in the z-direction.

Tangent Planes

The sphere $(x - 4)^2 + (y - 4)^2 + (z - 4)^2 = 16$ is centered at (4, 4, 4) and has radius $r = 4$. It is tangent to all three coordinate planes just as a beach ball pressed into the corner of a room is tangent to two walls and the floor. Planes such as these are called *tangent planes*. They are analogous to tangent lines in a plane.

There is nothing special about the coordinate planes in this example. There are planes tangent to the sphere at each point on its surface. (Just imagine sliding a stiff sheet of cardboard around the sphere. The cardboard represents a plane tangent to the sphere.)

Just as the derivative of a single-variable function gives us the information we need to find the equation of a line tangent to a curve, so too the partial derivatives provide information that allows us to find equations of tangent planes.

Example 7 Find the equation of the plane tangent to the paraboloid $z = x^2 + y^2$ at $(-3, 1, 10)$. The surfaces are shown in Figure 4.3.

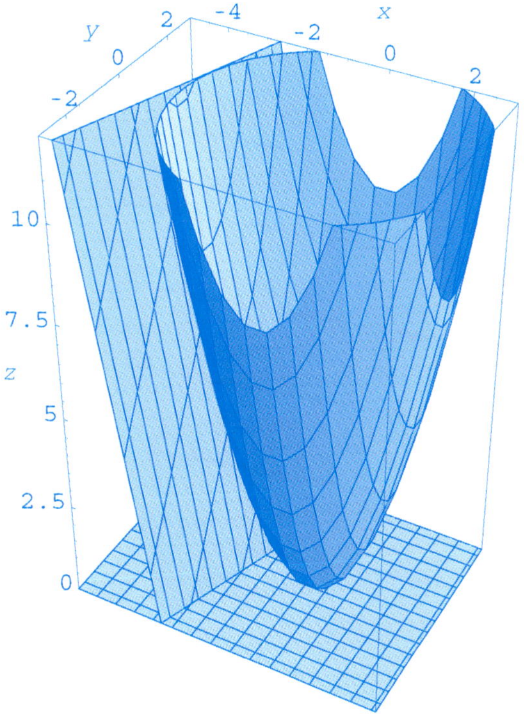

Figure 4.3

Solution We compute $\frac{\partial z}{\partial x} = 2x$ and $\frac{\partial z}{\partial y} = 2y$. At $(-3, 1, 10)$, $z_x = -6$ and $z_y = 2$.

Using the point-slope form of a plane, we can immediately write the solution as $z - 10 = -6(x + 3) + 2(y - 1)$.

Problems

1. **a.** Find the partial derivatives of $z = \sin(xy)$ and evaluate them at $(\pi, 2, 0)$.

   b. Is there any point on any x-section or any y-section at which the surface might have a vertical tangent plane?

2. **a.** Find f_x and f_y for the function

   $$f(x, y) = xe^{-x^2 - y^2}.$$

   b. Evaluate the partial derivatives at (0, 0) and (1, 1).

3. Figure 4.4 shows the level curves of a particular function.

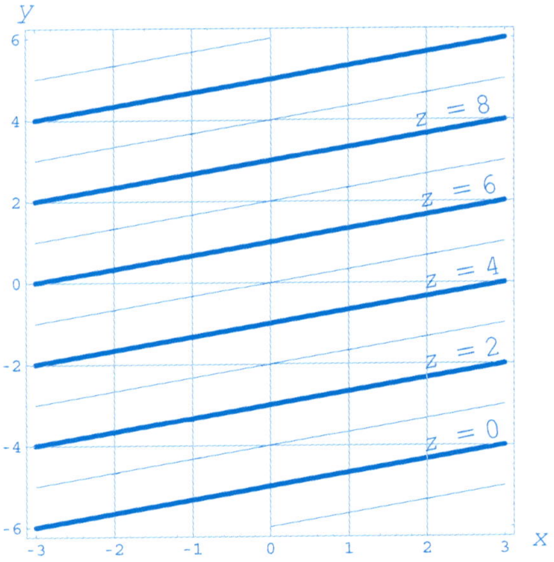

Figure 4.4

Use them to estimate the following partial derivatives.

   a. Estimate $\left.\frac{\partial z}{\partial x}\right|_{(2,\,2)}$ and $\left.\frac{\partial z}{\partial y}\right|_{(2,\,2)}$.

   b. Estimate $\left.\frac{\partial z}{\partial x}\right|_{(-1,\,0)}$ and $\left.\frac{\partial z}{\partial y}\right|_{(-1,\,0)}$.

4. Figure 4.5 shows level curves for some function $z = g(x, y)$.

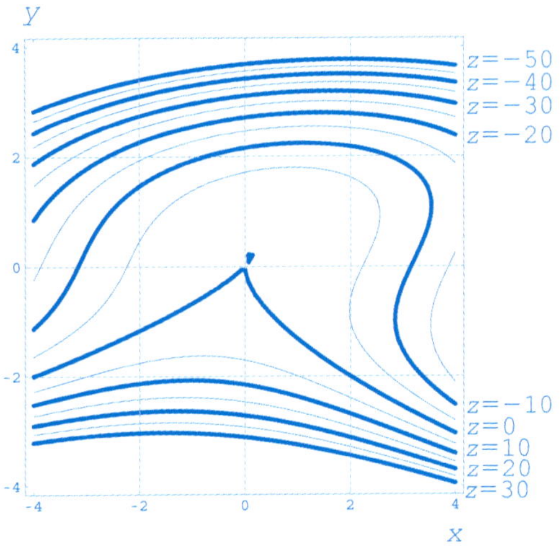

Figure 4.5

   a. What are the signs of g_x and g_y in each quadrant?

   b. Approximate $g_y(-2, 2)$ and $g_x(-2, 2)$.

5. This question is about the contour diagram for $z = f(x,y)$ shown in Figure 4.6

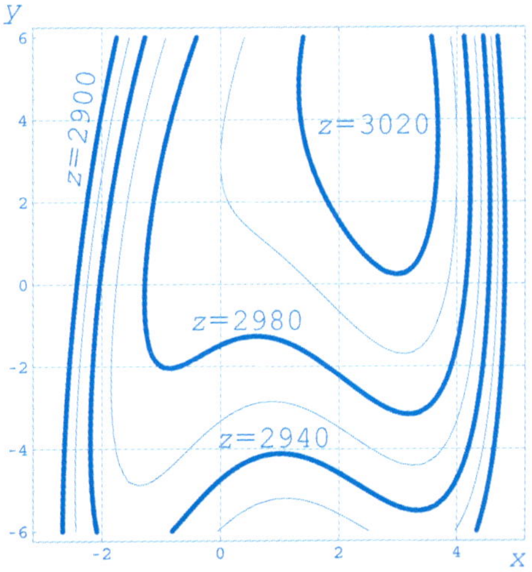

Figure 4.6

   a. Approximate $f_x(2,-4)$ and $f_y(2, -4)$.

   b. Approximate $f_x(-2, 3)$ and $f_y(-2, 3)$.

6. Table 4.3 below gives several values of a two-variable function. Is the function a plane? Explain your answer.

x \ y	0	1	2	3	4
0	2	−1	−4	−7	−10
1	4	1	−2	−5	−8
2	6	3	0	−3	−6
3	8	5	2	−1	−4
4	10	7	4	1	−2

Table 4.3

7. Table 4.4 below gives several values of a function $z = h(x, y)$. Use it to approximate $h_x(1, -1)$ and $h_y(-1, 1)$.

x \ y	−2	−1	0	1	2
−2	4	3	4	7	12
−1	3	1	1	5	7
0	4	1	0	2	4
1	7	3	1	1	3
2	12	9	4	3	4

Table 4.4

8. Find the equation of the tangent plane to $z = 2\sin\left(\frac{y}{3x}\right)$ at $\left(\frac{1}{\pi}, 1, \sqrt{3}\right)$.

9. Find the equation of the tangent plane to $z = \ln(x^2 - y^2)$ at $(2, 1, \ln(3))$.

10. Find the equation of a plane tangent to $z = 2\cos(x + y)$ when $x = y = \pi$.

11. a. Find the equation of a plane tangent to $z = |x| + |y|$ at $(1, -1, 2)$
    b. Find the equation of a plane tangent to $z = |x| + |y|$ at $(-2, 3, 5)$
    c. What is the equation of a tangent plane wherever x and y are both positive?
    d. What does the surface look like?

12. a. Find a function $f(x, y)$ satisfying the differential equation $\frac{\partial z}{\partial x} = 2xy$ if $z(0, 0) = 3$.
    b. Find a different function $g(x, y)$ satisfying the same partial differential equation with the same initial value if you also know that $\frac{\partial z}{\partial y} = x^2 - \cos(y)$.

Problem 12 illustrates a way in which differential equations with partial derivatives are more complex than ordinary differential equations. When you antidifferentiate, instead of adding an arbitrary constant, you must add an arbitrary function in the other variable. Use this idea to solve the partial differential equations in problems 13–15.

13. Find the general solution to $\frac{\partial z}{\partial y} = 4xy - 2y$ if you know also that $\frac{\partial z}{\partial x} = 2x + 2y^2$ and $z(2, 2) = 2$.

14. Find the general solution to $\frac{\partial z}{\partial x} = 4xy - 2y$ if you know also that $\frac{\partial z}{\partial y} = 2x^2 - 2x + ye^{y^2}$ and $z(2, 0) = 2$.

15. Find the general solution to

$$\frac{\partial z}{\partial x} = x\sqrt{x^2 + 2y} - 3$$

if you know also that

$$\frac{\partial z}{\partial y} = \sqrt{x^2 + 2y} + \sin(2y) + 1$$

and $z(-1, 0) = 0$.

The next two problems refer to the formal definition of partial derivatives as limits of difference quotients. Just as the ordinary derivative is defined as $\frac{dy}{dx} = \lim_{h \to \infty} \frac{f(x + h) - f(x)}{h}$ when the limit exists, so too are partial derivatives defined as limits of difference quotients. The partial derivative with respect to x is defined as $\frac{\partial z}{\partial x} = \lim_{h \to \infty} \frac{f(x + h, y) - f(x, y)}{h}$ where y is held constant as h approaches zero. Similarly, the partial derivative with respect to y is defined as

$$\frac{\partial z}{\partial y} = \lim_{h \to \infty} \frac{- f(x, y)}{}.$$

16. Use the limit definition of the derivative to find $\frac{\partial z}{\partial x}$ for the function $z = x^2 - \frac{1}{y}$.

17. Use the limit definition of the derivative to find $\frac{\partial z}{\partial y}$ for the function $z = |x| - 2y$.

CHAPTER 5

Surfaces of Revolution

In calculus we study solids of revolution, that is, solids formed by rotating an area in the xy-plane around a horizontal or vertical line. The surfaces of these solids are called *surfaces of revolution*. Since these surfaces are, of course, three-dimensional, they can be expressed as multivariable equations. In this chapter we will learn how to find equations for surfaces of revolution. Then, in the next two chapters, we will use our knowledge of how to find volumes of solids of revolution using single-variable calculus to examine volumes under any multivariable surface.

Example 1 When the portion of the parabola $y = 16 - x^2$, where $x \geq 0$, is rotated around the y-axis, it forms a solid of revolution. Find a multivariable equation for the surface of this solid.

Solution Figure 5.1 shows this surface and the plane $y = 2$. We will use sections to find the equation of this parabolic surface.

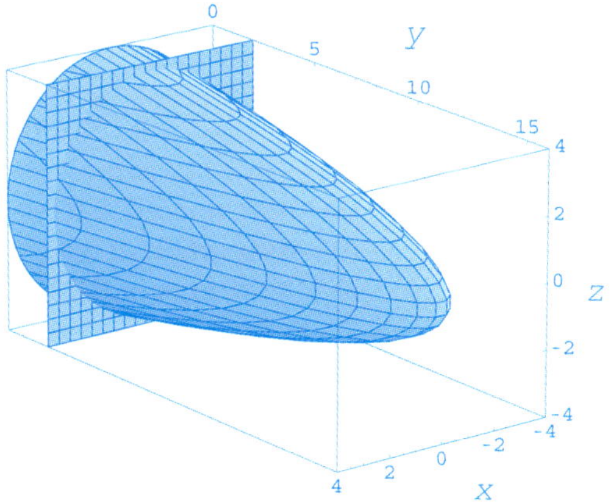

Figure 5.1

The equation of a surface must hold for each point on the surface, so we begin by considering a typical point (x_0, y_0) on the parabola. As the curve is rotated around the y-axis, this point rotates in a circle with radius $r = x_0$. Note that this circle lies in the y-section $y = y_0$. In Figure 5.1 we can see this circle when $y = 2$ as a particular y-section of the surface.

Since this intersection lies in an xz-plane, its equation will involve just those variables. In general, the equation of the circle in the y-section $y = y_0$ (which contains the point (x_0, y_0)) would be $x^2 + z^2 = x_0^2$. For example, the equation of the circle in the plane $y = 2$ is $x^2 + z^2 = 14$, since $x^2 = 14$ when $y = 2$. But this cannot be the equation of the surface, since it only holds in one particular y-section. We would like to replace the specific value x_0 with the general radius for these circles in any y-section. It would be an error to write $x^2 + z^2 = x^2$, however. You will have a chance to reason through this mistake in the problem set.

The correct way to get the general radius in terms of y is to use the original equation to solve for x^2, giving $x^2 = 16 - y$. Now substitute this general radius for the specific radius x_0 in the equation of the circular cross-section to get the equation of the surface of revolution $x^2 + z^2 = 16 - y$.

The reasoning we used to find the equation of this surface can, in some cases, be generalized. When the curve $y = f(x)$ is rotated around the y-axis, a typical point (x_0, y_0) rotates into the circle $x^2 + y^2 = y_0^2$ lying in the plane $y = y_0$. As we do this, we will assume that we are rotating a portion of $y = f(x)$ where $x \geq 0$ or where $x \leq 0$. This will avoid the complication of two different parts of the graph overlapping when rotated around an axis.

In Example 1, we used the equation of the original curve to solve for x in terms of y in order to determine the general radius for the circle. However, we can only do that if the original function is invertible. If it is not, then a horizontal line will intersect the curve more than once and the surface might intersect a horizontal plane in more than one concentric circle. The graph of $y = \cos(x)$, for example, intersects $y = 0.4$ infinitely many times. If the curve is rotated around the y-axis, the surface of revolution will intersect the plane $y = 0.4$ in infinitely many concentric rings.

Because these surfaces are significantly more complex, we will not attempt to find equations for them in this little book. Instead, we will restrict ourselves to surfaces that intersect a y-section in, at most, one circle. That is, we will restrict ourselves to functions that intersect each horizontal line at most once, in other words, invertible functions. The astute reader may have noticed that the parabola in Example 1 is not an invertible function; however, the portion of the parabola that we worked with in this problem, that is, the part to the right of the x-axis, is invertible.

If $y = f(x)$ is any invertible function whose graph is rotated around the y-axis, the general radius is $x = f^{-1}(y)$, and the equation of a general y-section is $x^2 + z^2 = (f^{-1}(y))^2$. Since this is true for all values of x, y, and z on the surface, this formula will give the equation of the surface of revolution.

Example 2 Find the equation of the surface of revolution when $g(x) = \sqrt{x}$ is rotated around the y-axis.

Solution Here, g is invertible and $g^{-1}(y) = y^2$, so the surface is $x^2 + z^2 = y^4$. This surface is shown in Figure 5.2.

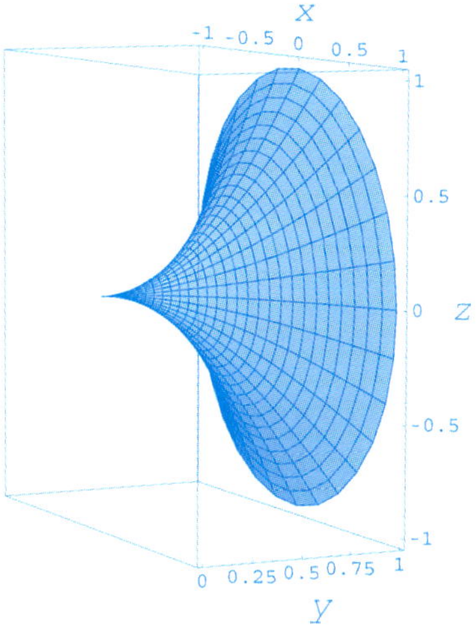

Figure 5.2

Example 3 Find the equation of the surface of revolution formed by rotating the line $y = \frac{x}{2}$ around the y-axis.

Solution The function $y = \frac{x}{2}$ is invertible and its inverse is $x = 2y$, so the equation of the surface is $x^2 + z^2 = 4y^2$. The surface is a pair of cones balanced tip-to-tip, as shown in Figure 5.3.

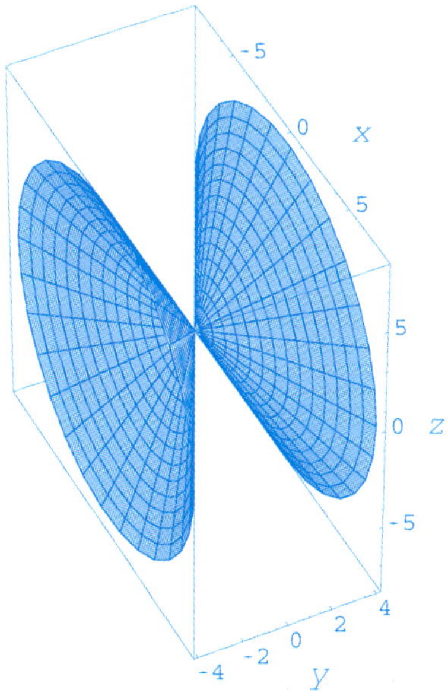

Now consider what happens if, instead, we rotate a function $y = f(x)$ around the x-axis. In this case, each x-section is a circle. Since the general radius of these x-section circles is $y = f(x)$, the equation of the surface is $y^2 + z^2 = f^2(x)$.

Example 4 Find the equation of the surface of revolution formed by rotating $y = \cos(x)$ around the x-axis.

Solution The intersection of this surface with any plane parallel to the yz-plane has radius $r = \cos(x)$. The equation of the cross-section is, therefore, $y^2 + z^2 = \cos^2(x)$. Part of the surface is shown in Figure 5.4.

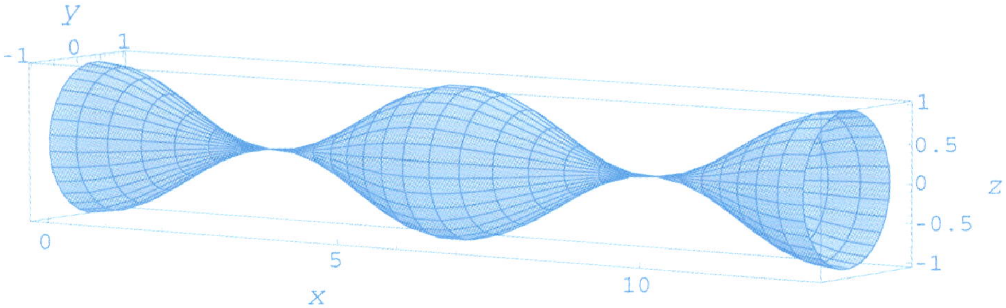

Figure 5.4

Notice that the formula for the general radius of a cross-section of a surface where the y-axis is the axis of rotation is expressed as a function of y, while the general radius of a circular cross-section of a surface formed by rotating a curve around the x-axis is a function of x. In general, the radius is expressed as a function of the variable on the axis of rotation.

Example 5 Figure 5.5a shows the surface of revolution $z = 2x^2 + 2y^2 - 18$. Find an equation of a curve that could be used to generate this surface.

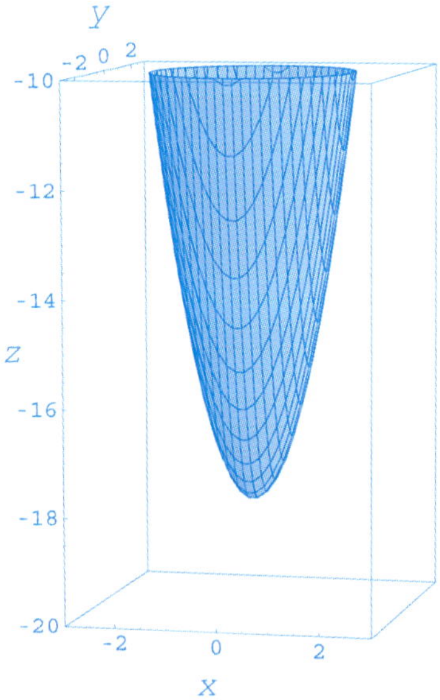

Figure 5.5a

Solution A typical level curve, say $z = 0$, is a circle centered at the origin. Therefore, the z-axis is the central axis of revolution and the curve we seek must have the form $z = f(x)$ or $z = f(y)$. To find the curve, slice the surface with a plane passing through the axis of symmetry, say $y = 0$, as shown in Figure 5.5b.

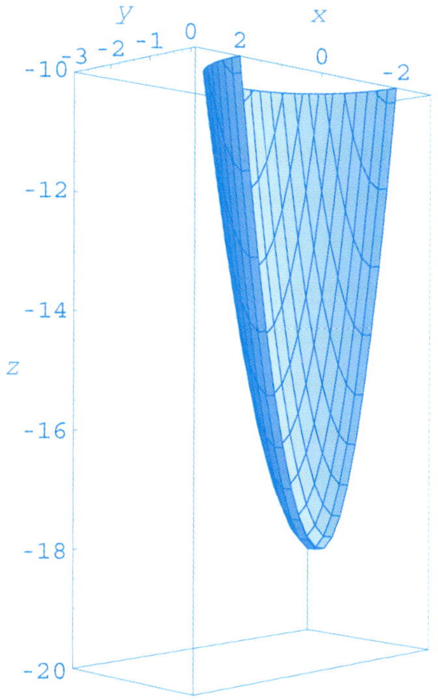

Figure 5.5b

Substitute in the original equation to get the equation of this y-section, the parabola $z = 2x^2 - 18$. Either half of this parabola (the part with $x > 0$ or the part with $x < 0$) will generate the entire surface when rotated around the z-axis.

If, instead, we had sliced the paraboloid with the plane $x = 0$, we would have gotten the x-section $z = 2y^2 - 18$. This parabola will also generate the surface when either half is rotated around the z-axis.

Example 6 Identify the surface $e^{-x} - y^2 - z^2 = 0$ in Figure 5.6 as a surface of revolution.

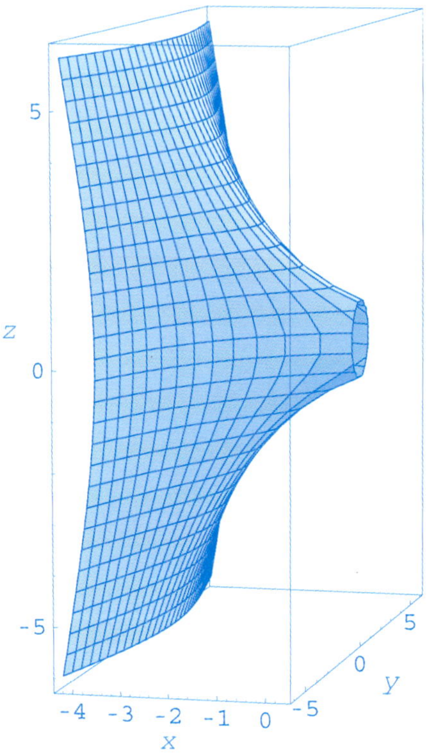

Figure 5.6

Solution Do you see the form of a circle lurking in this equation? The x-sections are circles $e^{-x} = y^2 + z^2$. These are centered on the x-axis and have the general radius $r = \sqrt{e^{-x}}$, suggesting that the surface can be formed by rotating a curve around the x-axis.

The surface is interesting because of its end behavior. As $x \to \infty$, the radius of a typical section decreases, approaching zero. In other words, the spike we see in Figure 5.6 is infinitely long.

To find a single-variable function which would rotate into this surface, slice the graph along the x-axis with the plane $y = 0$ (or with the plane $z = 0$). This gives the equation $e^{-x} = z^2$. If we solve for z, we get the equations $z = e^{\frac{-x}{2}}$ and $z = -e^{\frac{-x}{2}}$. Either of these will rotate into this surface. Had we chosen instead to take the z-section $z = 0$, we would get two other possibilities: $y = e^{\frac{-x}{2}}$ and $y = -e^{\frac{-x}{2}}$.

Problems

1. Show that the equation $y = 16 - x^2 - z^2$ gives the correct sections for the surface of revolution in Example 1.

2. Find the equation of the surface of revolution formed by rotating the parabola $y = 16 - 0.25x^2$ around the x-axis, shown in Figure 5.7. Notice that when the parabola is rotated around the x-axis that it forms three portions divided at the x-intercepts of the parabola.

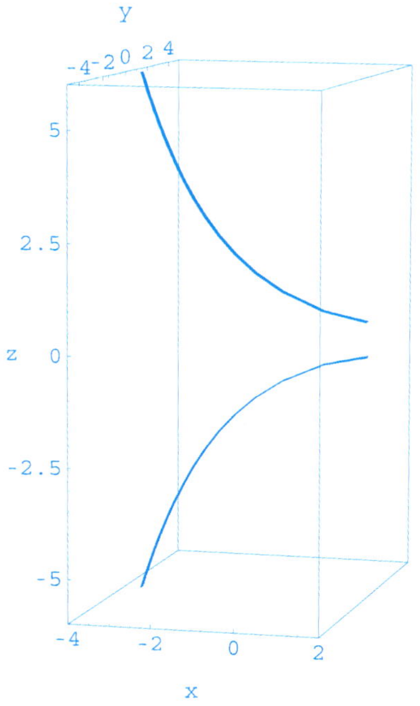

Figure 5.7

3. **a.** $y = \frac{1}{x^2}$ is rotated around the x-axis forming a surface of revolution. Find the equation of the cross-section of the surface with the plane $x = 4$.

   b. Find the equation of the surface.

   c. Find the equation of the surface of revolution formed when $y = \frac{1}{x^2}$ is rotated around the y-axis.

4. **a.** Find the equation of the surface of the cone formed by rotating the line $y = 2x$ around the x-axis.

   b. Find the equation of the surface of the cone formed by rotating the line $y = 2z$ around the x-axis.

5. **a.** Find the equation of the surface formed by rotating $y = 2^x$ around the y-axis.

   b. Find the equation of the surface formed by rotating $z = 2^y$ around the z-axis.

6. **a.** Find the equation of the surface of revolution formed by rotating the top half of the unit circle around the x-axis.

   b. Find the equation of the surface of revolution formed by rotating the portion of the unit circle in the first quadrant around the y-axis.

7. In Example 1, we rotated the parabola $y = 16 - x^2$ around the y-axis and found the equation of a typical circle in the plane $y = y_0$ was $x^2 + z^2 = x_0^2$. To move from this particular y-section to any y-section whose radius is an x-value of the parabola, we said that it would be a mistake to write $x^2 + z^2 = x^2$. Since this equation reduces to $z^2 = 0$, which obviously is not the correct equation of the surface, it is clear that this is a mistake. But why doesn't this reasoning work? That is, explain why using x for the radius, rather than $\sqrt{16 - y}$, is incorrect.

8. Find the volume enclosed by the surface $z - x^2 - y^2 = 0$ and the planes $z = 1$ and $z = 4$. (Don't let the unfamiliar z-variable bother you. This is just a solid of revolution and you know how to find their volumes.)

9. Find the volume enclosed by the surface $y = x^2 + z^2 - 4$ and the xz-plane as shown in Figure 5.8.

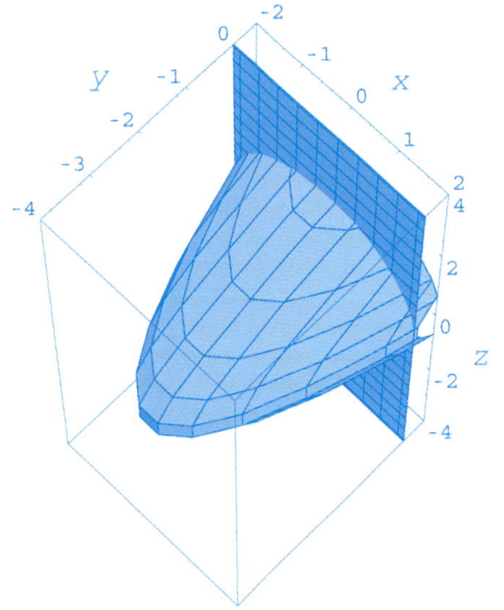

Figure 5.8

10. Figure 5.9 shows the top half of the solid of revolution $x^2 + y^2 = \frac{1}{z^2}$, which appears to be an infinitely long spike that quickly levels off to a flat surface.

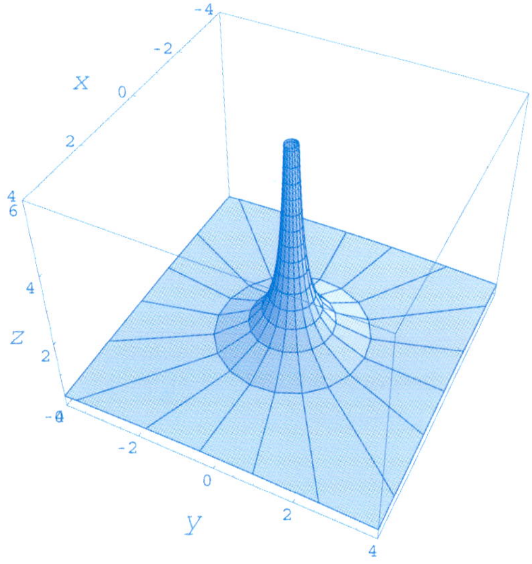

Figure 5.9

a. What is the equation of the asymptotic plane?

b. What is the central axis of the spike?

c. If you have studied improper integrals, you know that some unbounded solids can contain a finite volume. Assume that the figure is unbounded in the positive z-direction but limited, as shown, in the x- and y-directions to the circle of radius 4 shown in the figure. Find the volume enclosed by this solid of revolution.

11. Figure 5.10 shows the surface of revolution formed when the hyperbola $y^2 - x^2 = 1$ is rotated around the y-axis.

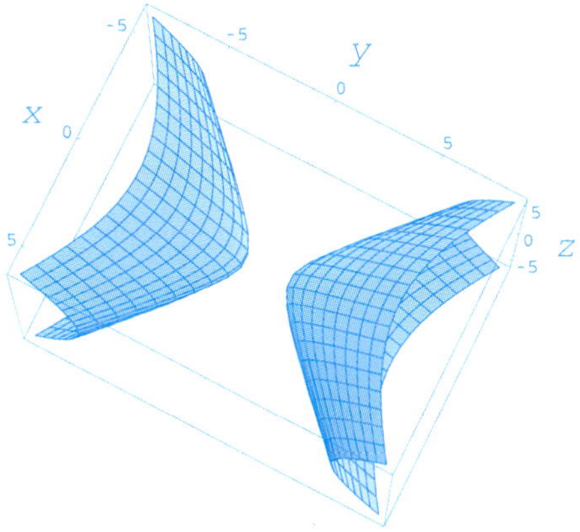

Figure 5.10

a. The surface is called a *hyperboloid of two sheets*. Find the volume inside the positive cup for $0 \le x \le 3$.

b. The hyperbola has two asymptotes, $y = x$ and $y = -x$. Describe the asymptotic behavior of the three-dimensional surface.

12. Figure 5.11 shows the surface of revolution formed when the hyperbola $y^2 - x^2 = 1$ is rotated around the x-axis.

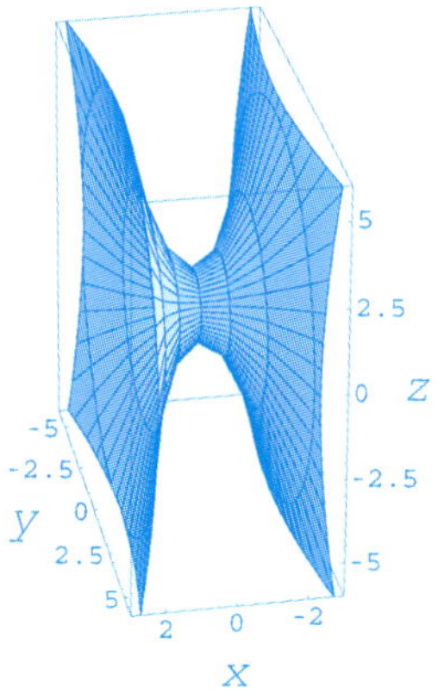

Figure 5.11

This is the same hyperbola as in Problem 10, but now there is a single continuous surface called a *hyperboloid of one sheet*. Find the volume inside the surface shown here for $-4 \le x \le 4$.

13. Figure 5.12 shows the plane $z = 9$ and the surface $z = (x - 2)^2 + (y - 4)^2$.

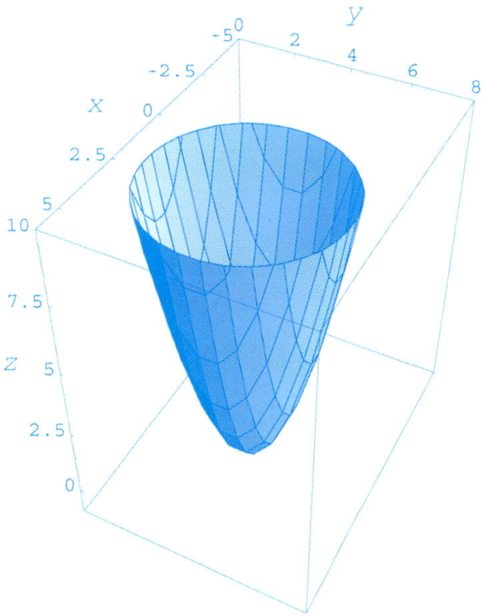

Figure 5.12

Find the volume enclosed by these surfaces.

14. Find the volume enclosed by the two paraboloids $z = 2x^2 + 2y^2 + 8$ and $z = x^2 + y^2 - 19$.

Hint: The volume is between the vertices of these two opposite facing paraboloids shown in Figure 5.13. Find the coordinates of their vertices and the equation of the plane where they intersect.

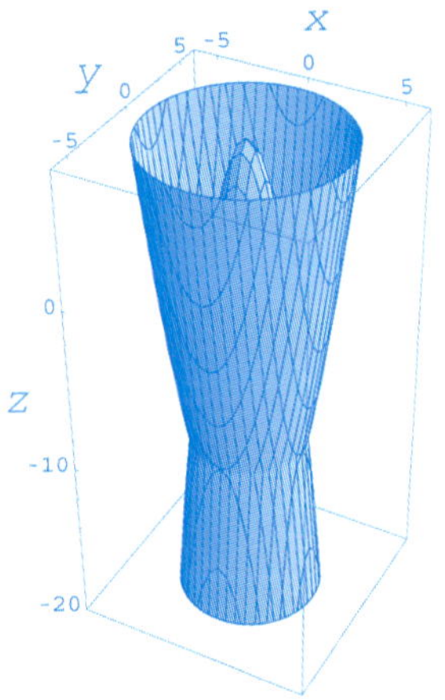

Figure 5.13

15. Find the volume enclosed by the two paraboloids $z = 2x^2 + 2y^2 + 26$ and $z = \frac{1}{2}x^2 + \frac{1}{2}y^2 + 38$ shown in Figure 5.14.

Hint: The volume is between two paraboloids that face in the same direction.

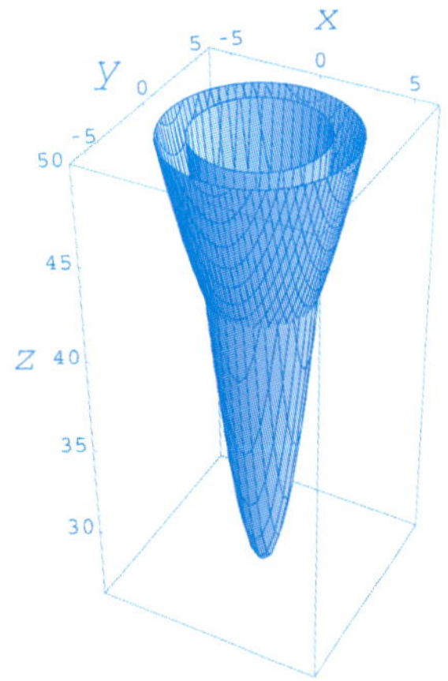

Figure 5.14

16. Find the volume shown in Figure 5.15, enclosed by the surface formed by rotating the top half of the ellipse $\frac{y^2}{16} + \frac{x^2}{9} = 1$ around the y-axis and the surface formed by rotating the bottom half of the ellipse $\frac{y^2}{100} + \frac{x^2}{9} = 1$ around the y-axis.
Hint: Treat the volume as the sum of two different volumes and set up two integrals to find the enclosed volume.

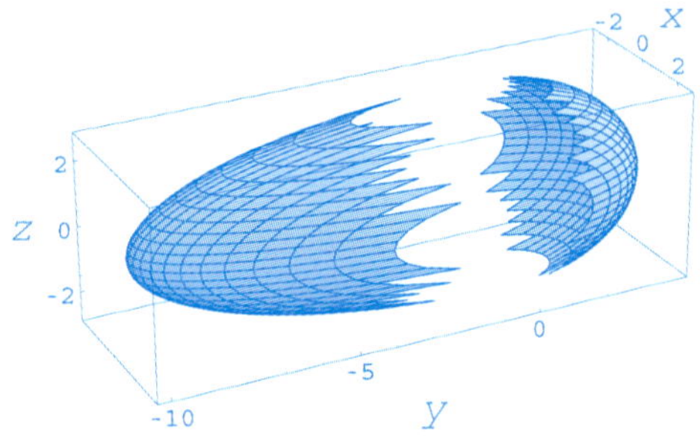

Figure 5.15

17. When the positive branch of $y = \frac{1}{x}$ is rotated around the x-axis, an infinitely long volume called Gabriel's Horn is created. A portion of this surface is shown in Figure 5.16.

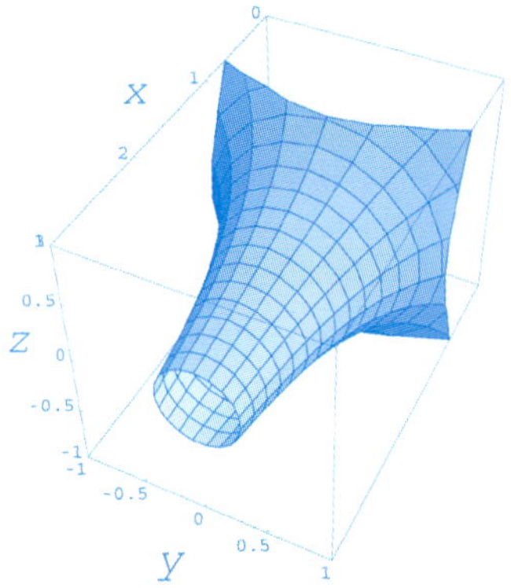

Figure 5.16

a. Find a multivariable formula for the surface of the Horn.

b. Find the volume enclosed by this surface where $x \geq 1$.

CHAPTER 6

Volume Under a Surface: Double Integrals

The problem of finding the volume under a surface, like the analogous problem of finding the area under a curve, is solved by integration. Of course, things are a bit more complicated in space.

In the previous chapter we examined solids of revolution. While the multivariable equations for the surfaces of these solids were new, the process of finding their volumes was not. In this chapter, however, we consider the problem of finding the volume between any surface and the xy-plane. We will need a different method when the volume is not formed by revolving a curve. The first example illustrates a process for approximating volume that will eventually lead to a new type of integral.

Since the problem of finding the volume under a surface is analogous to the problem of finding the area under a curve, let's begin by reviewing the procedure for finding the area under a curve. Computing the area between $y = e^{-x}$ and the x-axis for $0 \le x \le 2$, for example, should be a straightforward problem in integration for you, but it will be useful to recall the steps that led to the integral $\int_0^2 e^{-x}dx$.

Begin by dividing the interval [0, 2] into n parts each Δx wide. These slice the area into thin strips and the area of each strip can be approximated by a rectangle. Figure 6.1 illustrates this process with 8 rectangles so that $\Delta x = 0.25$. In each interval, we have chosen the left endpoint to establish the height of the rectangle. Note, therefore, that for each interval the area of the rectangle can be expressed as $e^{-x_k}\Delta x$, where x_k is the x-coordinate of the left endpoint of the interval.

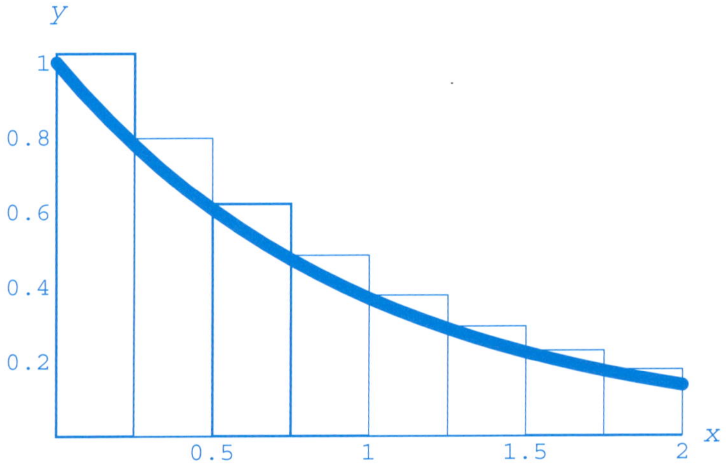

Figure 6.1

The sum of the areas of the rectangles approximates the area under the graph. As $\Delta x \to 0$, the Riemann Sum $\sum_{k=0}^{n-1}$ approaches the integral $\int_0^2 e^{-x}dx$, which gives us the area under the curve. We will use a similar strategy in Example 1 to approximate the volume under a surface.

Example 1 Approximate the volume between the surface $z = e^{-x} + 2e^{-y}$ and the square region in the xy-plane $0 \le x < 2$, $0 \le y \le 4$ shown in Figure 6.2.

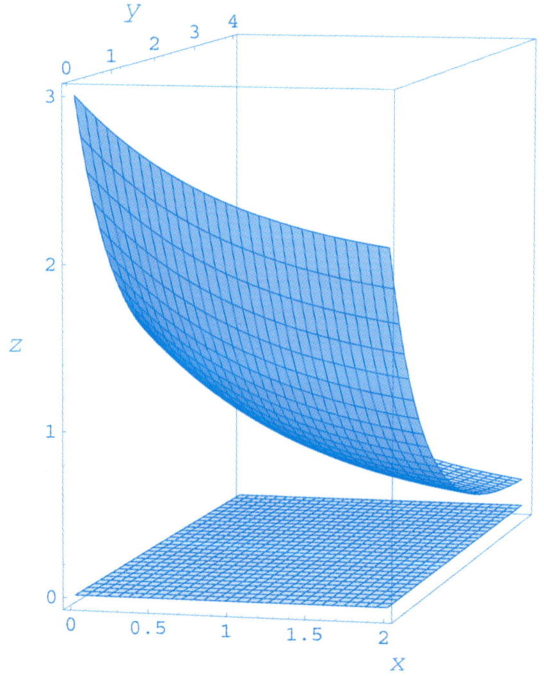

Figure 6.2

Solution

Subdivide the x-interval $[0, 2]$ into n parts of width Δx, and the y-interval $[0, 4]$ into m parts of width Δy. It isn't necessary that $\Delta x = \Delta y$ or that $m = n$. These partitions divide the $[0, 2] \times [0, 4]$ rectangle in the xy-plane into mn tiny rectangles each of which has an area of $\Delta A = \Delta x \Delta y$.

Now, imagine that the volume under the surface is filled with Jell-O™ and that the rectangular grid in the plane $z = 0$ is a wire screen. If we lift the screen upwards in the z-direction, it will slice the Jell-O into mn tall thin towers. Each tower has a rectangular base Δx long and Δy wide and a curved top formed by the surface $z = e^{-x} + 2e^{-y}$. The volume of each tower can be approximated by the volume of a rectangular box with the same base and with a height z chosen on the curved surface. The sum of the volumes of the boxes approximates the volume under the curve.

Table 6.1 gives the results of one particular partition. Specifically, we took $\Delta x = 0.5$ and $\Delta y = 1$, which divided the rectangle into 16 parts. In each small rectangle, the height of the approximating box was computed at the lower-left hand corner of the base. The values in the table are the volumes of each tower $\Delta V = (e^{-x} + 2e^{-y})(0.5)(1.0)$.

x \ y	0	1	2	3
0	1.5	0.86788	0.63534	0.54979
0.5	1.3033	0.67114	0.43860	0.35305
1.0	1.1839	0.55182	0.31928	0.23373
1.5	1.1116	0.47944	0.24690	0.16135

Table 6.1

The approximate volume under the curve using this specific partition is the sum of the 16 numbers in the table, so $V \approx 10.607$.

Approximating volumes this way is not fun, because there is a lot of computation and no easy shortcut. We used a spreadsheet to compute the values in the box and this helped somewhat, but the computations still take time. On the other hand, the method is basically familiar and, theoretically, it could produce a volume approximation to any degree of accuracy.

Just as rectangle approximations of area lead to Riemann sums that, in the limit, became definite integrals, these rectangular boxes are used in multivariable calculus to construct sums, which, in the limit, become *double integrals.*

In multivariable calculus, we form Riemann-type sums $\sum f(x, y)\Delta A = \sum f(x, y)\Delta x \Delta y$ where the summation is taken over all the x-values and all the y-values. We then take the limit as Δx and Δy each approach zero. When the limit exists it is called a double integral and is written as $\iint_A f(x, y)\,dA$.

These exotic integrals go well beyond the scope of this little book. If you would like to know more about them, you can consult a text on multivariable calculus or, better yet, continue studying calculus in college. For our purposes, however, we will deal with volumes like the one in this example by using a technique you are already familiar with from calculus: the method of slabs.

Recall that, in the method of slabs, volumes are sliced into thin slabs with similar cross-sections. When you used this method before, the cross-sections were always squares or triangles or circles, figures whose area you could find with geometry. Now, however, we will allow the cross-sections to be any area we can compute by integration.

In Figure 6.3a, the surface $z = e^{-x} + 2e^{-y}$ has been sliced by two planes parallel to the yz-plane and Δx apart.

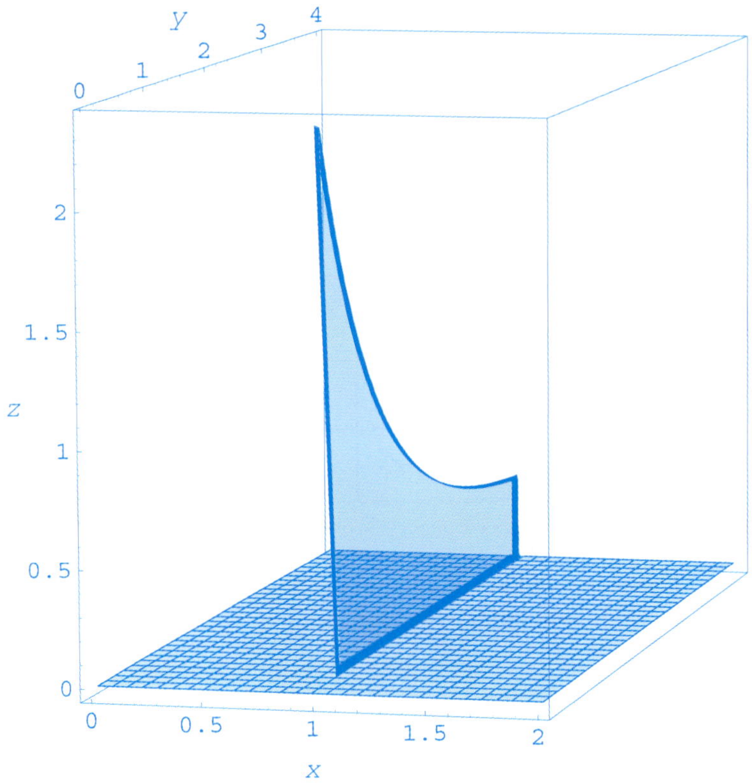

Figure 6.3a

These planes form a thin slice of the volume under the surface. These slices are formed by partitioning the interval $0 \le x \le 2$ and slicing with planes $x = x_i$ at each partition point. This partition slices the volume into thin slabs of width Δx. In Figure 6.3a, we show a typical slab formed by slicing a figure at a typical value we call $x = c$ and at a nearby value $x = c + \Delta x$.

In Figure 6.3b, we have altered our point of view so that we are looking down on the yz-plane. From this direction the front of the slab looks like the area under a graph, the x-slice $z = e^{-x} + 2e^{-y}$(where x is held constant).

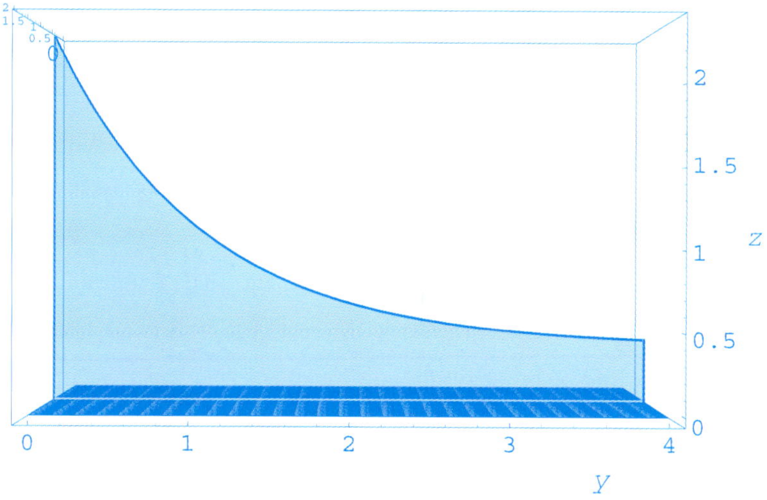

Figure 6.3b

The volume of this slab is approximately the volume of a prism whose height is Δx and whose base area is the area under the curve $z = e^{-c} + 2e^{-y}$ on the interval $0 \le y \le 4$. This area can be computed with an integral:

$A = \int_0^4 (e^{-c} + 2e^{-y})\,dy = (e^{-c}y - 2e^{-y})\Big|_0^4 = 4e^{-c} - 2e^{-4} + 2$. The volume of a slab with this cross-section is, approximately $\Delta V \approx (4e^{-c} - 2e^{-4} + 2)\Delta x$.

The method of slabs gives the volume under a surface as $V = \int_a^b A(x)dx$, where $A(x)$ is the area of a typical cross-section. In our example, we used $x = c$, but, in fact, the area we found can be written as a function of the position x in the interval. That is,

$A(x) = 4e^{-x} - 2e^{-4} + 2$. The total volume will therefore be

$V = \int_0^2 (4e^{-x} - 2e^{-4} + 2)dx = (-4e^{-x} - 2e^{-4}x + 2x)\Big|_0^2 = -4e^{-2} - 4e^{-4} + 8 \approx 7.385.$

Notice that our approximation of 10.607 was an overestimate and not very accurate. To have obtained 7.385 as our estimate would have required partitioning the grid into tens of thousands of subdivisions. Clearly, approximation by towers is not a practical way to estimate the volume. Using the method of slabs to evaluate two integrals, however, is not much more work than evaluating an ordinary integral.

A Closer Look at the Method of Slabs

It is worth looking back at the process we used to compute a volume. We did not pull a rabbit out of a hat, but we did do something unusual. First we visualized the volume as a sum of volumes of approximating slabs of cross-section $A(x)$. That is, for each value of x in the interval $0 \leq x \leq 2$, we computed the area of a cross-section. The cross-section itself was the area under a curve *in the y-direction*. That area was computed as a dy integral. The result was a formula for the cross-sectional area *as a function of x*. The final step was to integrate this function over the x-interval.

We could streamline the process in the following way. To find the volume under the surface $z = e^{-x} + 2e^{-y}$ over the region $0 \leq x \leq 2,\ 0 \leq y \leq 4$, first compute $\int_0^4 (e^{-x} + 2e^{-y})dy$ while treating x as a constant. The result will be a function of x. Then, perform a second integration with respect to x. It is customary to combine the two processes into a single symbol and write $V = \int_0^2 \int_0^4 (e^{-x} + 2e^{-y})dy\, dx$. This is called an *iterated integral* and is computed by doing two separate integrations, first with respect to y (while holding x constant) and then with respect to x.

The idea of holding x constant while integrating is similar to the idea of holding x constant while differentiating. The latter gave us the concept of a partial derivative. Perhaps you could think of the inner integral in an iterated integral as a partial integral.

In general, the volume under a surface can be expressed as a double integral. In practical terms, however, we try to evaluate the double integral as an iterated integral.

Example 2 The surface $y = x^2 + \sin(2x + z) + 1$ lies above the plane $y = 0$. Find a double integral for the volume between this surface and the xz-plane over the boundaries $0 \leq x \leq \frac{\pi}{2}$ and $0 \leq z \leq \frac{3\pi}{2}$ and then evaluate it as an iterated integral.

Solution The volume between the surface $y = x^2 + \sin(2x + z) + 1$ and the xz-plane is given by the double integral $\iint_A (x^2 + \sin(2x + z) + 1)dA$, where A is the rectangle $[0, \frac{\pi}{2}] \times [0, \frac{3\pi}{2}]$. We will compute the volume as an iterated integral.

But we have two choices. We can compute the volume by slicing in the z-direction or in the x-direction. If we slice in the z-direction, then the area of a typica slice (for a fixed value of z) is $A = \int_0^{\frac{\pi}{2}} (x^2 + \sin(2x + z) + 1)dx$ and then these areas can be added up in the z-direction to determine the volume by evaluating the iterated integral $V = \int_0^{\frac{3\pi}{2}} \int_0^{\frac{\pi}{2}} (x^2 + \sin(2x + z) + 1)dx\, dz$. Alternatively, we could add up the slices formed by slicing in the x-direction. The area of a typical slice in this case would be $A = \int_0^{\frac{3\pi}{2}} (x^2 + \sin(2x + z) + 1)dz$. We then would compute the volume by summing these slabs using the iterated integral $\int_0^{\frac{\pi}{2}} \int_0^{\frac{3\pi}{2}} (x^2 + \sin(2x + z) + 1)\, dzdx$. Does the order matter?

We will compute the first integral here and leave the second one for the problems at the end of the chapter and you will then be able to decide for yourself whether the order of integration matters. We begin evaluating the integral

$V = \int_0^{\frac{3\pi}{2}} \int_0^{\frac{\pi}{2}} (x^2 + \sin(2x + z) + 1)dx\, dz$ by computing the inner integral.

$\int_0^{\frac{\pi}{2}} (x^2 + \sin(2x + z) + 1)dx =$

$(\frac{x^3}{3} - \frac{1}{2}\cos(2x + z) + x)\Big|_0^{\frac{\pi}{2}} = \frac{\pi^3}{24} - \frac{1}{2}\cos(\pi + z) + \frac{\pi}{2} + \frac{1}{2}\cos(z).$

Notice that this integral is a function of z. That is because it represents the area of a typical slice, and the area depends upon which z-section we select.

Next, to find the volume we use the method of slabs. That is, $V = \int_0^{3\pi/2} A(z)dz$, so that the volume under the surface will be

$$V = \int_0^{\frac{3\pi}{2}} \left(\frac{\pi^3}{24} - \frac{1}{2}\cos\,(\pi + z) + \frac{\pi}{2} + \frac{1}{2}\cos(z)\right)dz \text{ or}$$

$$V = \left(\frac{\pi^3 z}{24} - \frac{1}{2}\,\sin(\pi + z) + \frac{\pi z}{2} + \frac{1}{2}\sin(z)\right)\Bigg|_0^{\frac{3\pi}{2}} = \frac{\pi^4}{16} - \frac{1}{2}\sin\left(\frac{5\pi}{2}\right) + \frac{3\pi^2}{4} + \frac{1}{2}\sin\left(\frac{3\pi}{2}\right) \text{ or}$$

$$V = \frac{\pi^4}{16} + \frac{3\pi^2}{4} - 1.$$

Once again, when we use an iterated integral to evaluate a double integral as in this example, there will always be a choice: should we evaluate the integral as $dA = dx\,dy$ or as $dA = dy\,dx$? In Example 2, we chose to slice the volume using z-slices and the resulting integral had the form $dx\,dz$. This volume could also have been computed using x-slices. The result would be an iterated integral with dz on the inside and dx on the outside. In Problem 8, you will be asked to evaluate the iterated integral in this form, and you will see for yourself if the order of integration in an iterated integral makes a difference. Since both integrals are supposed to compute the same volume, you might suspect that the order is unimportant.

There is certainly a sense in which the order of integration is critical, however. Sometimes one order will lead to integrals that are easy to compute while the reverse order will lead to integrals that are difficult or even impossible to evaluate. When you have a choice you need to decide whether one order is easier than the other. Experience is the best teacher here. The more problems you do, the better you will become at recognizing which direction to choose. For now, the best guideline is this: if slicing in one direction results in an integral too difficult to evaluate, try slicing in the other direction.

Problems

1. Compute $\int_0^{\pi}\int_0^{\frac{\pi}{2}} 3\sin(x+z)dx\,dz$

2. Compute $\int_0^1\int_0^1 \frac{y}{1+x^2}dy\,dx$

3. Figure 6.4 gives a view of the part of the surface $z = \cos(x) + \cos(y)$ above the xy-plane. This part of the surface lies above the rectangle in the xy-plane whose corners lie at $(\frac{-\pi}{2}, \frac{\pi}{2})$, $(\frac{-\pi}{2}, \frac{-\pi}{2})$, $(\frac{\pi}{2}, \frac{\pi}{2})$, and $(\frac{\pi}{2}, \frac{-\pi}{2})$. Set up and evaluate an iterated integral to find the volume between this surface and the xy-plane.

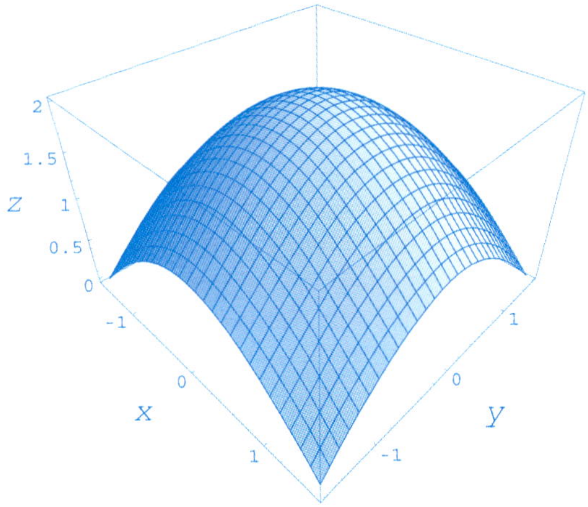

Figure 6.4

4. Compute $\int_{-1}^1\int_0^4 z^3\sqrt{3x+4}\,dx\,dz$. Can you explain the result?

5. Find the volume under the surface $z = e^{x+y}$ and above the xy-plane for $0 \le x \le 2$ and $0 \le y \le 4$

6. The box formed by the planes $|x-4| = 4$, $|y-5| = 5$, and $|z-2| = 2$ has dimensions $8 \times 10 \times 4$. Compute the volume of 320 in three different ways, by writing and evaluating three different iterated integrals, one with the variables x and y, a second with the variables y and z, and a third with the variables z and x.

7. Use an iterated integral to confirm the volume of the hemisphere $z = \sqrt{r^2 - x^2 - y^2}$ is $\frac{2}{3}\pi r^3$.

8. The volume in Example 2 was computed by evaluating the integral $V = \int_0^{\frac{\pi}{2}}\int_0^{\frac{3\pi}{2}} (x^2 + \sin(2x+z) + 1)dx\,dz.$ Now, compute it by reversing the order of the integration and demonstrate that the result is the same. That is, show that $V = \int_0^{\frac{\pi}{2}}\int_0^{\frac{3\pi}{2}} (x^2 + \sin(2x+z) + 1)dz\,dx = \frac{\pi^4}{16} + \frac{3\pi^2}{4} - 1.$

CHAPTER 7

Average Value of a Function

If $y = f(x)$ is a single-variable function, its average value over an interval $[a, b]$ is defined to be $\frac{1}{b-a}\int_a^b f(x)dx$. The average value can be thought of as the (signed) area between the curve and the x-axis distributed over the length of the interval of integration. Similarly, we can define the *average value* of a multivariable function $z = g(x, y)$ as $\frac{1}{A}\iint_A g(x,y)dA$ where we distribute the (signed) volume between the surface and the region over the area A of the region of integration.

Example 1 Find the average value of the function $z = \sqrt{r^2 - x^2 - y^2}$ over its domain.

Solution The function is the surface of a hemisphere of radius r. The domain of the function is the disc $x^2 + y^2 = r^2$. The average value is $\frac{1}{\pi r^2}\iint_A z\, dA$ since the area is $A = \pi r^2$. As we saw in problem 7 of Chapter 6, the double integral computes to $\left(\frac{2}{3}\right)\pi r^3$, the volume of a hemisphere of radius r, so the average value is $z = \frac{\left(\frac{2}{3}\right)\pi r^3}{\pi r^2} = \frac{2r}{3}$.

We can interpret this value in several ways. It can be viewed, for example, as the average height above the xy-plane of a bug randomly crawling on the surface of this hemisphere. It is also the height of a cylinder of radius r with the same volume as the hemisphere.

This example did not require anything more than geometry to evaluate the quantities involved. The next example, however, requires the evaluation of a double integral.

Example 2 Find the average value of $z = \sin(x) + \cos(y)$ over the square $[0, \pi] \times [0, \pi]$.

Solution Since the area of the region is π^2, the average value of the function is

$$\frac{1}{\pi^2}\int_0^{\pi}\int_0^{\pi}(\sin(x) + \cos(y))dx\,dy = \frac{1}{\pi^2}\int_0^{\pi}(-\cos(x) + x\cos(y))\Big|_0^{\pi} dy$$

$$= \frac{1}{\pi^2}\int_0^{\pi}(2 + \pi\cos(y))dy = \frac{1}{\pi^2}(2y + \pi\sin(y))\Big|_0^{\pi} = \frac{2}{\pi}.$$

Example 3 Find the average value of the function $z = x^a y^b$ over the unit square.

Solution The average value is just the volume between the surface and the xy-plane since the area of the square under the surface is 1. We again compute the double integral as an iterated integral

$$\int_0^1\int_0^1 x^a y^b dx\,dy = \int_0^1 \frac{x^{a+1}y^b}{a+1}\Big|_0^1 dy = \frac{1}{a+1}\int_0^1 y^b\,dy = \frac{1}{(a+1)(b+1)}.$$

Example 4 The plane $z = 12 - 3x - 2y$ intersects the coordinate axes at (4, 0, 0), (0, 6, 0), and (0, 0, 12). Together with the coordinate planes, this forms a pyramid whose base is a right triangle, as shown in Figure 7.1. Find the average value of this function over that right triangle.

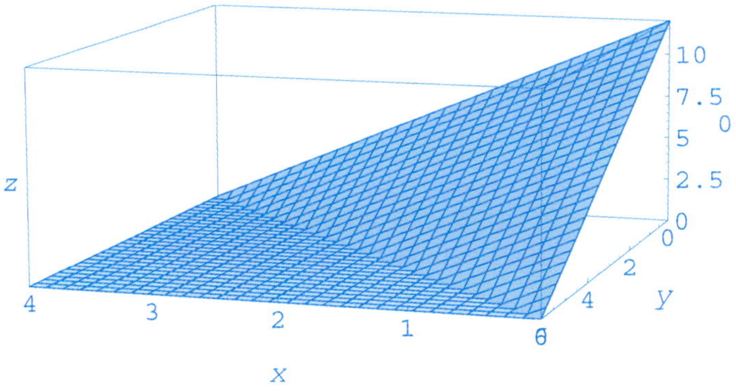

Figure 7.1

Solution The base of the pyramid is a right triangle whose legs are 4 and 6, so the area of the base is 12. The average value will be computed from the formula $\frac{1}{12}\iint_A (12 - 3x - 2y)dA$. This is another integral that is easier to evaluate geometrically. The double integral represents the volume of the pyramid, which is $V = \frac{1}{3}hA$ where h is the pyramid's height and A is the area of the base. Thus, the average value is $\frac{V}{A} = \dfrac{\frac{1}{3}hA}{A} = \frac{1}{3}h = \frac{1}{3}(12) = 4.$

Example 5 Find the average value of the function $f(x, y) = e^{0.6(x + y)}$ above the right triangle bounded by the x and y-axes and the line $y = 4 - 2x$.

Solution The intercepts of $y = 4 - 2x$ are (0, 4) and (2, 0), so the area of the base (in the plane $z = 0$) is 4. The region is shown in Figure 7.2a. The average value of f is, therefore, $\frac{1}{4}\iint_A e^{0.6(x + y)}dA$. The problem is how to write the double integral as an iterated integral. When the domain is a rectangle, as in Examples 2 and 3, it is easy to write an iterated integral since the limits of x and y are intervals. However, in this example, the domain of x is $0 \le x \le 2$ and the domain of y is $0 \le y \le 4$, but it would not be correct to write $\int_0^4\int_0^2 e^{0.6(x + y)}dxdy$. Do you see why? This iterated integral is evaluated over a rectangle since x takes on all of its values for each value of y, but that is not what is happening in this example. The inner integral represents the area of a slice perpendicular to the y-axis. Figure 7.2b shows just the base of a typical y-slice. Notice that, in this slice, x does not take on all of its values from 0 to 2 but only those values between the plane $x = 0$ and the plane $y = 4 - 2x$. This plane can be rewritten as $x = \frac{4 - y}{2}$, so the values of x in this slice are those in the interval $0 \le x \le \frac{4 - y}{2}$.

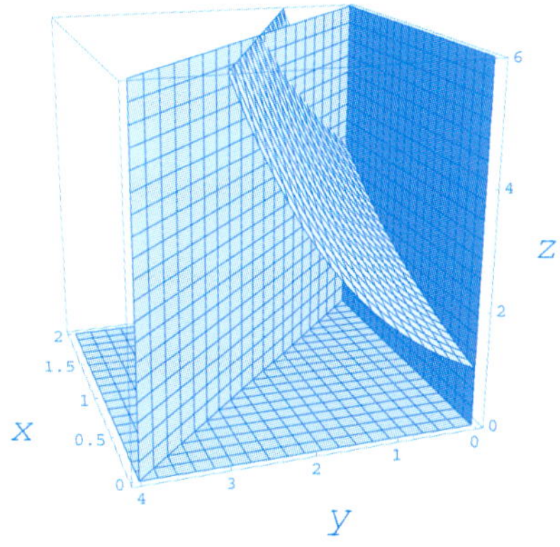

Figure 7.2a

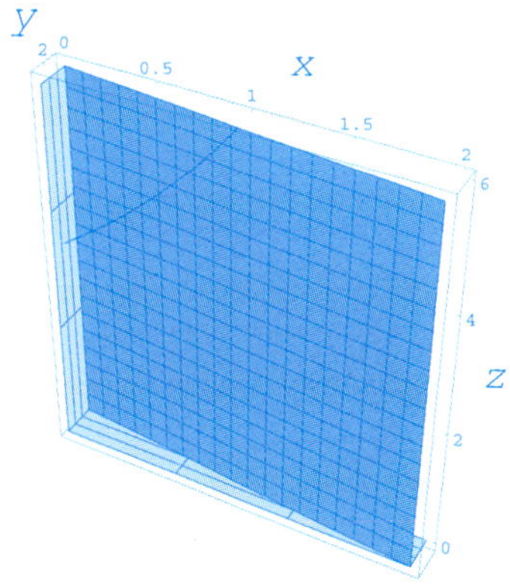

Figure 7.2b

In order to represent the altering widths of the slabs, the inner integral must be written with a variable upper limit. For each choice of y in the domain [0, 4], x varies from $x = 0$ to $x = \frac{(4 - y)}{2}$. The integral $\int_0^{\frac{(4 - y)}{2}} e^{0.6(x + y)}dx$ gives the volume of any particular y-slab. The integral looks formidable, but it is not quite as bad as it looks. Remember that when we compute the inner integral we treat y as constant. The antiderivative of $e^{0.6(x + y)}$ is simply $\int e^{0.6(x + y)}dx = \frac{1}{0.6}e^{0.6(x + y)} + C$, and so the inner integral is computed as $\int_0^{\frac{(4 - y)}{2}} e^{0.6(x + y)}dx = \frac{1}{0.6}e^{0.6(x + y)}\Big|_0^{\frac{(4 - y)}{2}} = \frac{5}{3}\left(e^{\frac{0.6(4 + y)}{2}} - e^{0.6y}\right)$.

Therefore, $\int_0^4 \int_0^{\frac{(4-y)}{2}} e^{0.6(x+y)}dx\,dy =$

$$\frac{5}{3}\int_0^4 \left(e^{\frac{0.6(4+y)}{2}} - e^{0.6}\right)dy = \frac{5}{3}\left(\frac{2e^{0.3(4+y)}}{0.6} - \frac{e^{0.6y}}{0.6}\right)\Bigg|_0^4 = \frac{5}{3}\left(\frac{2e^{2.4}}{0.6} - \frac{e^{2.4}}{0.6}\right) - \frac{5}{3}\left(\frac{2e^{1.2}}{0.6} - \frac{e^{0}}{0.6}\right).$$

Finally, $\frac{5}{3}\left(\frac{e^{2.4}}{0.6}\right) - \frac{5}{3}\left(\frac{2e^{1.2}-1}{0.6}\right) \approx 14.953$. Therefore, the average value is $\frac{14.953}{4} = 3.738$.

Iterated Integrals Over Nonrectangular Regions

Example 5 illustrated the complexity of double integrals when the domain is not a simple rectangle. As another example, Figure 7.3 shows the *elliptical paraboloid* $z = 9 - x^2 - 9y^2$. How can we compute the volume between this surface and the xy-plane?

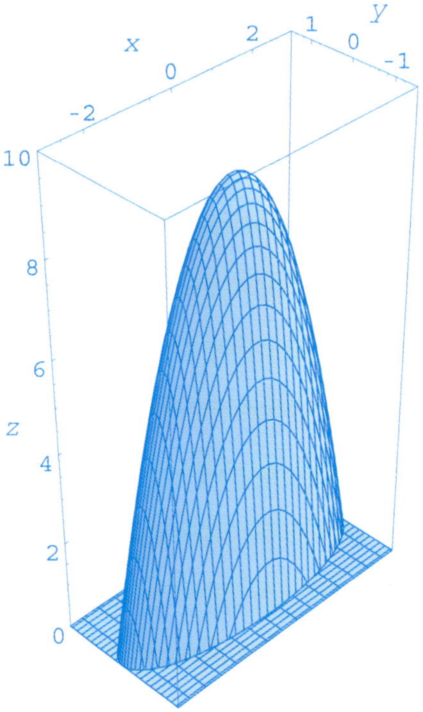

Figure 7.3

The x- and y-sections of this thin lollipop shape are parabolas, while the z-sections are ellipses, so this is not a surface of revolution and the surface does not extend over the full rectangle $-3 \le x \le 3$, $-1 \le y \le 1$. The volume is given by the double integral $\iint_A (9 - x^2 - 9y^2)dA$ where A is the area of the ellipse $x^2 + 9y^2 = 9$ in the xy-plane.

There are many ways to compute the value of this double integral. The simplest is to avoid multivariable calculus and slice the solid parallel to the xy-plane into elliptical slabs. You will try this as an exercise in the problem set. Alternatively, we can slice the solid parallel to one of the other coordinate planes and write the double integral as an iterated integral across its elliptical base. We will illustrate one of these solutions here and leave the reverse iteration as an exercise.

Example 6 Write the double integral $\iint_A (9 - x^2 - 9y^2)dA$, where A is the area of the ellipse $x^2 + 9y^2 = 9$ in the xy-plane, as an iterated integral. Evaluate the integral.

Solution A typical x-slice is shown in Figure 7.4.

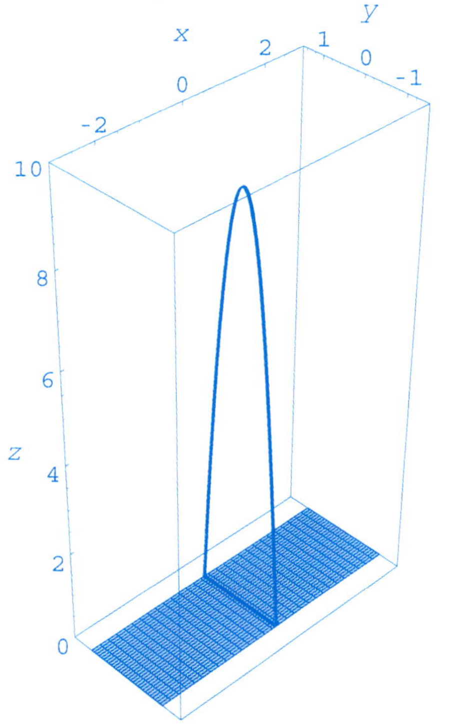

Figure 7.4

Notice that the parabolic slice does not quite reach the two sides of the box. Instead, the bottom of the parabola, in the plane $z = 0$, rests on the boundaries of the ellipse $0 = 9 - x^2 - 9y^2$, which we get by setting $z = 0$ in the original equation. A typical x-slice treats x as constant. We can find the y-values on this parabola by solving for y^2 in the equation of the ellipse. In the plane $z = 0$, the equation $y^2 = \frac{1}{9}(9 - x^2)$ gives the two symmetrical y-values for each x-value on the ellipse. They are $y = -\frac{1}{3}\sqrt{9 - x^2}$ and $y = \frac{1}{3}\sqrt{9 - x^2}$. The inner integral, therefore, ranges over the interval $-\frac{1}{3}\sqrt{9 - x^2} \le y \le \frac{1}{3}\sqrt{9 - x^2}$, so that the iterated integral is $\int_{-3}^{3}\int_{-\frac{1}{3}\sqrt{9-x^2}}^{\frac{1}{3}\sqrt{9-x^2}} (9 - x^2 - 9y^2)dy\,dx$. This is messy, but the mess is just algebra; the hard part of the problem is over. An antiderivative of the inner integral is $(9 - x^2)y - 3y^3$ (remember, we are treating x as constant). Therefore, the inner integral evaluates to

$\left((9 - x^2)y - 3y^3\right)\Big|_{-\frac{1}{3}\sqrt{9-x^2}}^{\frac{1}{3}\sqrt{9-x^2}} = \frac{2}{3}(9 - x^2)^{\frac{3}{2}} - \frac{2}{9}(9 - x^2)^{\frac{3}{2}} = \frac{4}{9}(9 - x^2)^{\frac{3}{2}}$, and so the iterated integral becomes $\frac{4}{9}\int_{-3}^{3}(9 - x^2)^{\frac{3}{2}}dx \approx 42.412$.

Problems

1. Find the average value of the function $z = x\sin(y) + y\sin(x)$ over $[0, \frac{\pi}{2}] \times [0, \frac{\pi}{2}]$.

2. Find the average value of the function $f(x, y) = \frac{xy}{x^2 + y^2}$ over $[1, 2] \times [1, 2]$.

3. Find the average value of the function $g(x, y) = \frac{x^2 + y^2}{xy}$ over $[1, 2] \times [1, 2]$.

4. Find the average value of the function $z = xye^{x^2 + y^2}$ over $[0, 1] \times [0, 1]$.

5. **a.** The rectangle in the xy-plane with corners at (0, 0), (2, 0), (2, 3), and (0, 3) is the bottom of a box whose slanted top is a portion of the plane $z = 12 - 2x$. Use geometry to find the average height of this box with a slanted top as shown in Figure 7.5.

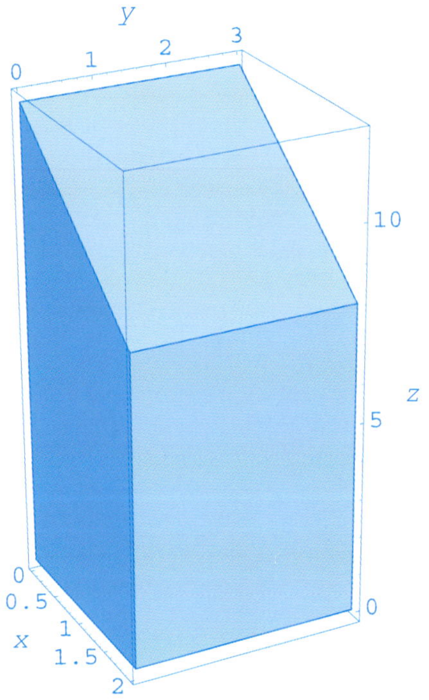

Figure 7.5

b. Find the average height of the box by setting up an iterated integral and evaluating it.

6. **a.** The plane $z = f(x, y)$ whose intercepts are (5, 0, 0), (0, 8, 0), and (0, 0, 12) forms a right pyramid with the three coordinate planes. Use geometry to show that the average value of this function is $z = 4$.

b. Find the average height of this pyramid by setting up an iterated integral and evaluating it.

7. The cost of manufacturing x frilbies and y classops is given by

$C = 0.2x^2 - .0033xy + 0.1y^2 + 18.$

Find the average cost if

$500 \leq x \leq 1000$ and $300 \leq y \leq 700$.

8. Figure 7.6 shows a triangle containing a 60° angle.

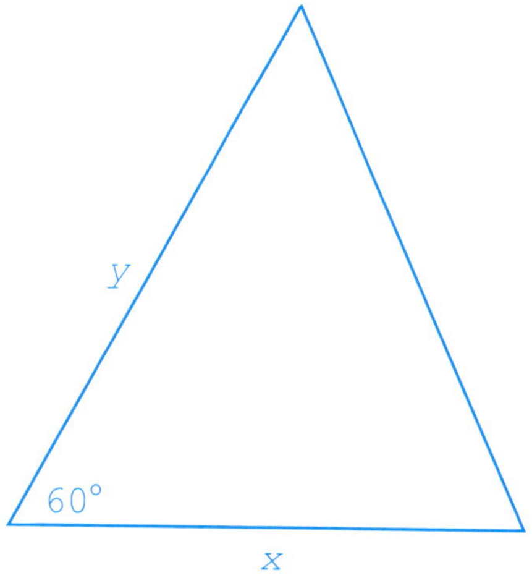

Figure 7.6

As x and y vary, the area of the triangle changes. Find the average area of triangles formed by varying x and y over the intervals $0 \leq x \leq 4$ and $0 \leq y \leq 8$.

Hint: The area of a triangle is $A = \frac{1}{2}xy\sin\theta$, where θ is the angle formed by the side of length x and the side of length y.

9. Figure 7.7 shows a box with a slanted top. The bottom of the box is the rectangle whose corners are (0, 0), (2, 0), (2, 3), and (3, 0), while the slanted top is a portion of the plane $z = 9 - \frac{x}{2} - y$. Find the average height of the box by setting up an iterated integral and evaluating it.

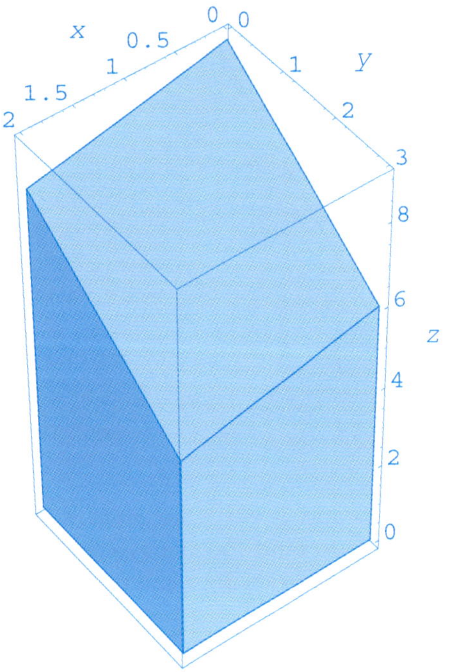

Figure 7.7

10. In Example 6, we wrote the double integral $\iint_A (9 - x^2 - 9y^2)dA$, where A is the ellipse $x^2 + 9y^2 = 9$, as an iterated integral

$$\int_{-3}^{3}\int_{-\frac{1}{3}\sqrt{9-x^2}}^{\frac{1}{3}\sqrt{9-x^2}} (9 - x^2 - 9y^2)dy\,dx.$$

a. Write the double integral as an iterated integral with $dx\,dy$ and compute its exact value.

b. Compute the exact value of the double integral by slicing the solid parallel to the plane $z = 0$ and using the method of slabs to compute the exact volume.

c. Find the average value of z over the region A, the area of the elliptical base.

11. Evaluate each of the following iterated integrals:

a. $\int_0^4\int_{2y}^3 (x + y)dx\,dy$

b. $\int_0^{\sqrt{x\backslash 2}}\int_0^x x^2\cos(xy)dy\,dx$

c. $\int_0^1\int_{-x}^x x^2y^3dy\,dx$

12. Consider the function $z = x^2y$. Set up, but do not evaluate, an iterated integral representing the volume between this surface and the xy-plane over each of the regions described.

a. The region above the x-axis and below the upper half of the unit circle.

b. The region enclosed by $y = x^2$ and $y = 6 - x$.

c. The region enclosed by the x-axis, the y-axis, and the line $y = 2x + 4$.

13. Find the volume above the unit circle and below the plane $z = 3x + 2y + 4$.

CHAPTER 8

Review Problems and Self-Assessment

1. Find a set of inequalities that describes the inside of the box in Chapter 7, Figure 7.7.

2. Find the coordinates of a point on the surface $z = x^2 + 2y^2 - 4$ which is

   a. below the xy-plane b. behind the xz-plane c. in the plane $x + y = 2$.

3. What are the coordinates of the corners of the box defined by the inequalities

   $|x - 4| \leq 7,\ |y + 3| \leq 2,\ |z - 6| \leq 8$?

4. Find the equation of the surface whose x and y-sections are shown in Figure 8.1a and 8.1b, respectively.

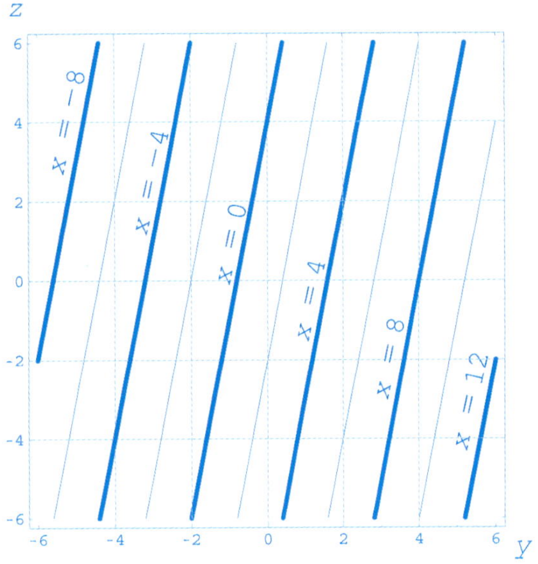

Figure 8.1a

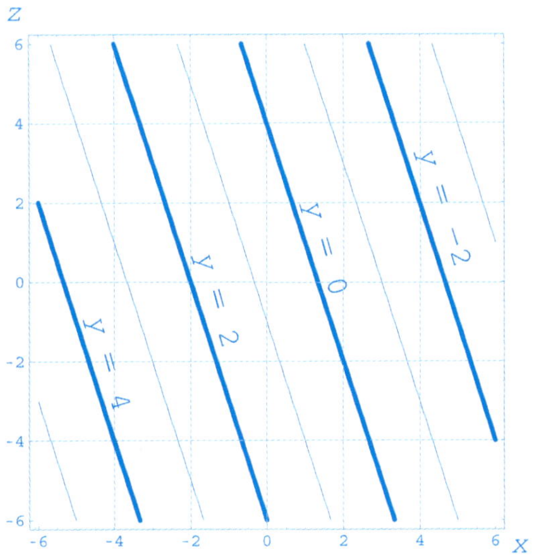

Figure 8.1b

5. Consider the surface $z = xy^2$.

a. Describe the x-sections.

b. Describe the y-sections.

6. The x and y-sections of a surface are shown in Figure 8.2a and 8.2b. Sketch the surface and describe the z-contours of the surface.

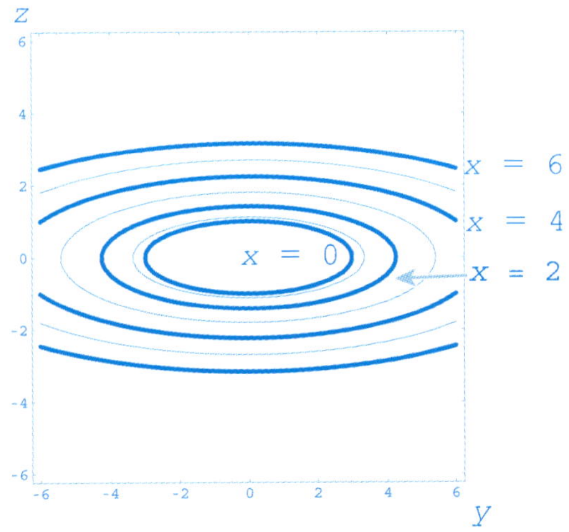

Figure 8.2a

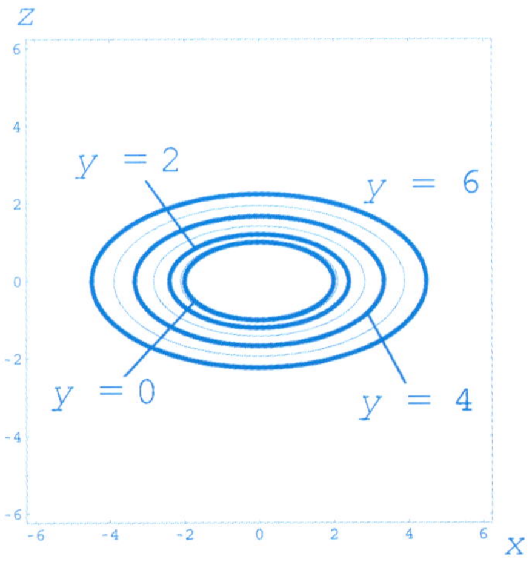

Figure 8.2b

7. Figure 8.3 shows a sinusoidal surface. Describe its x- and y- sections.

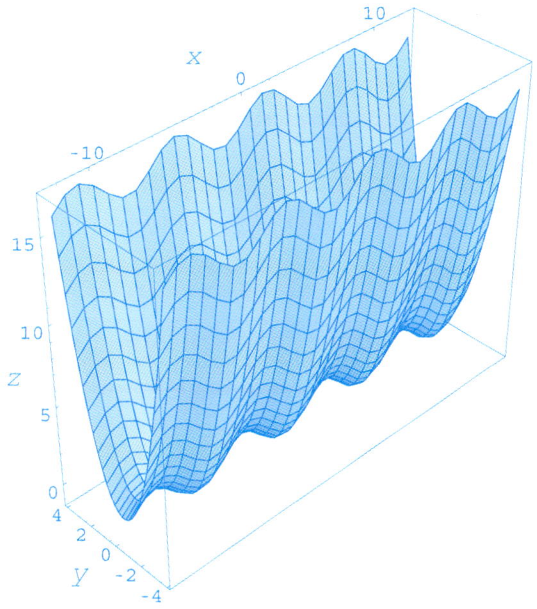

Figure 8.3

8. Figures 8.4a and 8.4b, respectively, show the contours and the x-sections of a certain multivariable function. Sketch the surface.

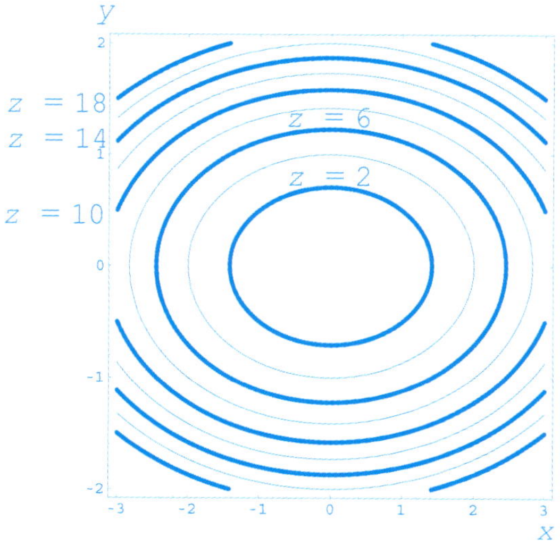

Figure 8.4a

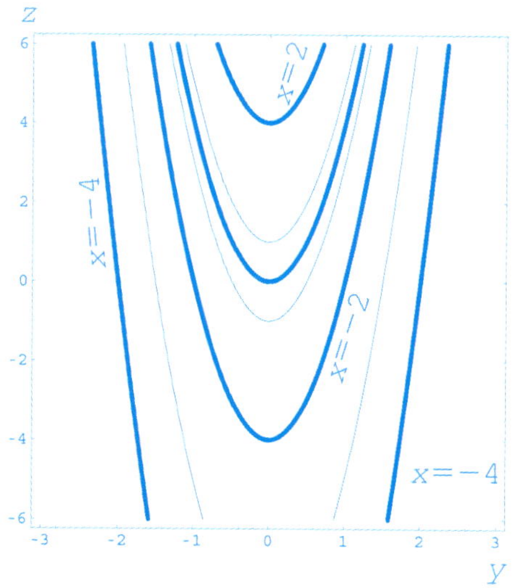

Figure 8.4b

9. Find a formula for a multivariable function whose level curves are ellipses parallel to the yz-plane.

10. Find $\frac{\partial z}{\partial x}$ and $\frac{\partial z}{\partial y}$ for each of the following functions.

a. $z = x^3y^2$ **b.** $z = \sin(x + y) - \cos(x - y)$ **c.** $z = e^{xy}\ln(x + y)$

11. If $\frac{\partial z}{\partial x} = -3\sin(x)\sin(y)$ and $\frac{\partial z}{\partial y} = 3\cos(x)\cos(y)$, find a formula for the multivariable function $z = f(x,y)$ that passes through the origin.

12. **a.** Use the contour diagram in Figure 8.5 to approximate $f_x\left(\frac{1}{4}, -\frac{1}{2}\right)$ and $f_y\left(-1, \frac{3}{2}\right)$.

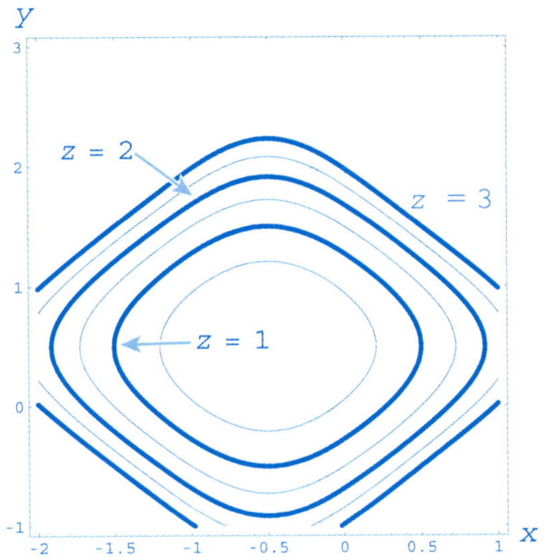

Figure 8.5

b. If you are at $(-0.5, 0.5)$, are you inside a valley or on top of a hill?

13. **a.** Find the equation of the surface formed by rotating $y = \ln(x)$ around the y-axis.

b. Find the volume enclosed by the surface that you found in part **a.** The plane $y = \ln(2)$, and the xz-plane.

c. Since the y-axis is a vertical asymptote of $y = \ln(x)$, the volume under $y = \ln(2)$ is unbounded. Express the volume as an improper integral and compute its value if the integral converges.

14. **a.** Find the equation of the surface formed by rotating the line $x = 3$ around the y-axis.

b. The volume enclosed by the surface in **a.** and the planes $y = 3$ and $y = 8$ is the volume of a cylinder of height 5 and radius 3. Set up an integral to compute this volume.

c. Find the equation of the cylindrical surface formed when $x = 3$ is rotated around $x = 5$.

15. **a.** Find the equation of the surface formed by rotating $y = x^{\frac{1}{3}} - 1$ around the y-axis.

b. Find the equation of the surface formed by rotating $y = x^{\frac{1}{3}} - 1$ around the x-axis.

c. Find the volume enclosed by the surface in **b.** and the planes $|x| = 8$.

16. **a.** Set up a double integral for the volume under the surface $z = \sqrt{x\sin(y)\cos(y)}$ and above the rectangle $0 \leq x \leq 4$, $0 \leq y \leq \frac{\pi}{2}$ in the xy-plane.

b. Write the integral in **a.** as an iterated integral and compute the volume.

17. **a.** Set up a double integral for the volume under the surface $z = x^3 + y^3$ and above the rectangle $-4 \leq x \leq 4$, $-4 \leq y \leq 4$ in the xy-plane.

b. Show that $V = \int_{-4}^{4} (x^3 + y^3)dx\,dy = 0$ and so cannot represent the volume.

c. Find the correct iterated integral for the volume and evaluate it.

18. **a.** Set up a double integral for the volume under the surface $z = e^{xy} + xy$ and above the rectangle $0 \leq x \leq 2$, $1 \leq y \leq 3$ in the plane $z = 0$.

b. Write the integral in **a.** as an iterated integral and compute the volume.

19. Find the volume indicated in Figure 8.6, above the surface $z = x^4 + y^2$ and under the disc $x^2 + y^2 = 4$ in the xy-plane.

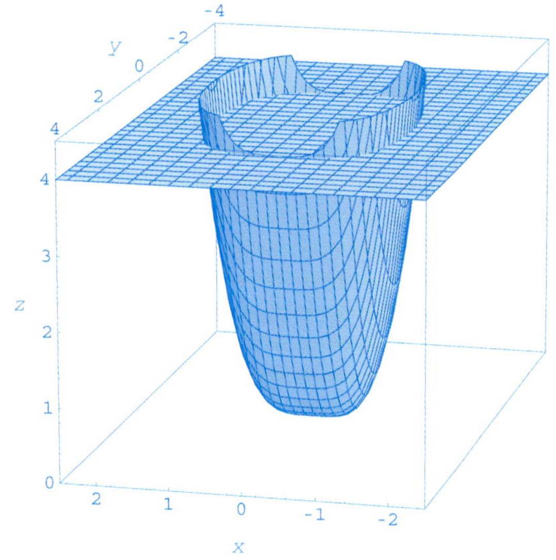

Figure 8.6

20. Find the volume under the surface $z = y^4 + x^2$ and above the triangle formed by the x- and y-axes and the line $2x + y = 4$, as indicated in Figure 8.7.

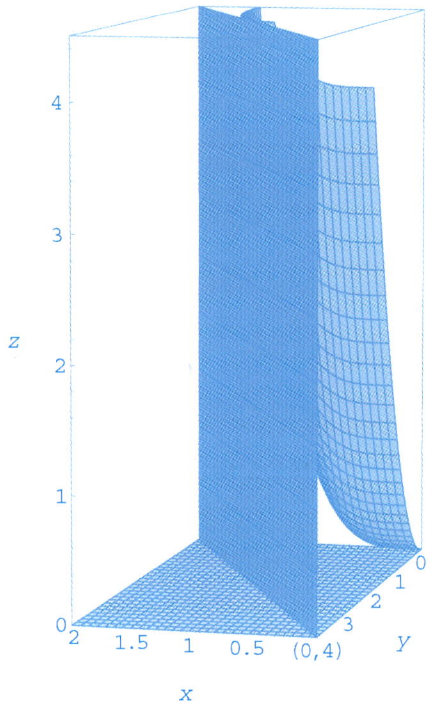

Figure 8.7

Self-Assessment

1. Name three points on the intersection of the cylinder $x^2 + z^2 = 25$ and the sphere $x^2 + y^2 + z^2 = 36$.

2. Find the equation of the plane passing through (1, 0, 1), (2, 1, 0), and (1, 1, 2).

3. Figures 8.8a and 8.8b, respectively, show x- and y-sections of some surface $z = f(x, y)$. A bug at (2, 3, 6) starts to crawl directly towards the origin. Will the bug be able to get there or will the surface be in the way?

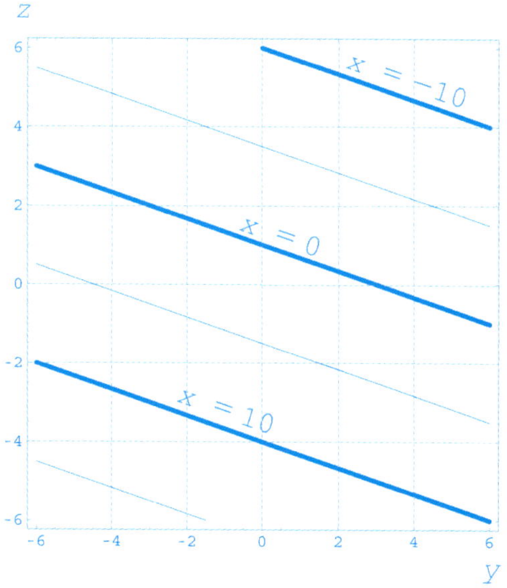

Figure 8.8a

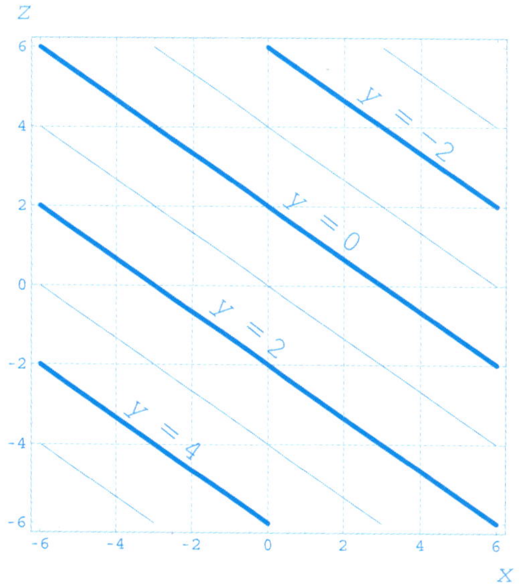

Figure 8.8b

4. A rectangular box is defined by the inequalities $|x - 6| \leq 8$, $|y + 10| \leq 2$, and $|z - 2| \leq 4$. Find the equation of a sphere whose center is at the center of the box and whose diameter is the length of the space diagonal of the box.

5. Figure 8.9 shows a contour diagram for some surface $z = g(x, y)$.

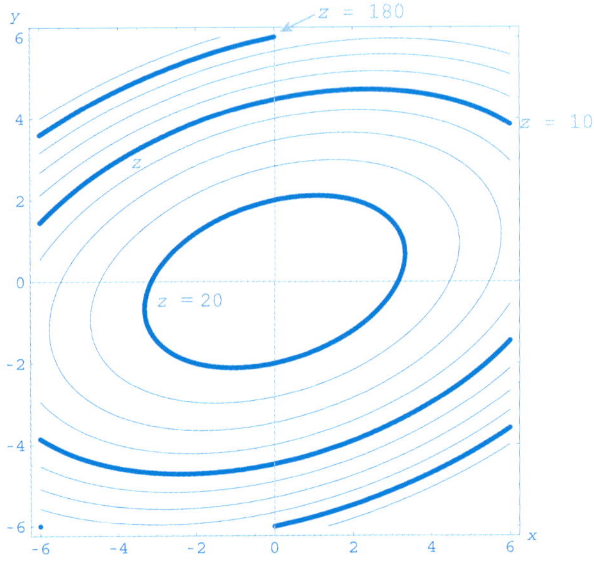

Figure 8.9

a. In which quadrants is $\frac{\partial z}{\partial x}$ always positive?

b. In which quadrants is $\frac{\partial z}{\partial y}$ always positive?

6. If $h(x, y) = e^{\sin(x + y)}$, evaluate $h_x(0, \pi)$ and $h_y\left(\frac{\pi}{12}, \frac{\pi}{12}\right)$.

7. **a.** The curve $y = \sqrt{x}$ is rotated around the y-axis. Find an equation for the surface of revolution.

b. The curve $y = \sqrt{x}$ is rotated around the x-axis. Find an equation for the surface of revolution.

8. **a.** Find the volume between the surface $x = 49 - z^2 - y^2$ and the yz-plane.

b. Find the average value of the function over the rectangle $-3 \leq y \leq 3$ by $-4 \leq z \leq 4$.

9. Find the volume enclosed by the paraboloids $z = 12 - (x - 1)^2 - (y - 5)^2$ and $z = \frac{1}{4}[(x - 1)^2 + (y - 5)^2] - 8$.

10. Evaluate a double integral to find the volume under the surface $z = xy$ and above the triangle in the xy-plane whose corners are (4, 4), (2, 4) and (4,2).

Answers to Odd-Numbered Problems and Self-Assessment

Chapter 1

1. Just as (2, –8) is 2 units from the y-axis and 8 units from the x-axis, the distance to each of the coordinate planes is the absolute value of one of the coordinates. So, the distance to the xy-plane is 2, to the yz-plane is 3, and to the xz-plane is 5.

3. The spheres are centered at (2, 5, –10) and (7, 1, –7). The distance between these centers is
$\sqrt{(2-7)^2 + (5-1)^2 + (-10+7)^2}$
$= \sqrt{50} = 5\sqrt{2}$.

5. $z = 0$

7. **a.** The central axis of the cylinder is the z-axis.
   b. The ordinary circle $x^2 + y^2 = 25$ lies in the xy-plane. Thus, any point on the cylinder that is not in the plane $z = 0$ will not lie on the ordinary circle $x^2 + y^2 = 25$. Such points can be described as (x, y, z), where (x, y) satisfy $x^2 + y^2 = 25$ and $z \neq 0$. For example, (–3, 4, 18) and (5, 0, 12).
   c. On the circle through (a, b, c), $a^2 + b^2 = 25$. Therefore, $(a, -b, c)$, $(-a, b, c)$ and $(-a, -b, c)$ all lie on the same circle.
   d. Since the axis of the cylinder is the z-axis, all points on the line passing through (a, b, c) have the same x- and y-coordinates but a different z-coordinate. For example, $(a, b, c + 1)$ and $(a, b, c - 12)$ both lie on the cylinder and on the line through (a, b, c).

9. The distance to the x-axis is the distance from (3, –5, 2) to (3, 0, 0) which is $\sqrt{5^2 + 2^2} = \sqrt{29}$. The distance to the y-axis is $\sqrt{13}$, and the distance to the z-axis is $\sqrt{34}$.

11. The inside of the cube lies between the parallel planes $x = -2$ and $x = 2$, between the planes $y = -2$ and $y = 2$, and between the planes $z = -2$ and $z = 2$. Therefore, the interior of the box can be described by the inequalities $|x| < 2$, $|y| < 2$, $|z| < 2$.

13. The x-coordinate has the smallest absolute value, so (3, –8, 4) is nearest the yz-plane. The equation of sphere centered at (3, –8, 4) with radius 3 is
$(x - 3)^2 + (y + 8)^2 + (z - 4)^2 = 9$.

15. Figure S1.15 shows that the solid is a triangular prism.

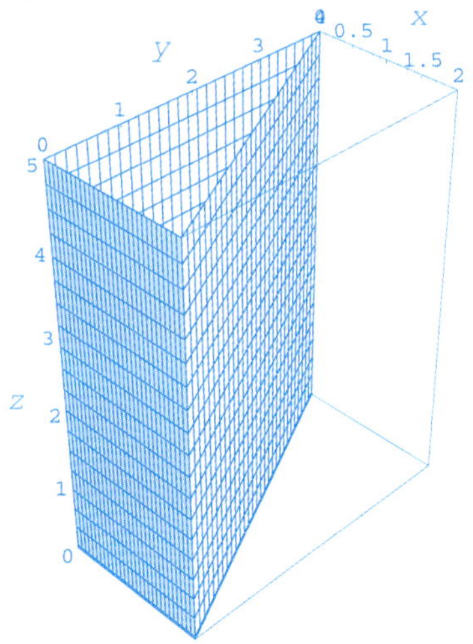

Figure S1.15

Note that the top of the prism is not shown so as to improve the view. Since the base is a right triangle whose sides are 2 and 4 and whose height is 7, the volume is $V = \frac{1}{2}(2)(4)(7) = 28$.

17. **a.** This is a sphere centered at (0, 1, 2) with radius 7 so its volume is $\frac{4}{3}\pi(7)^3 = \frac{1372\pi}{3}$.

b. When $y = 1$, the intersection is $x^2 + (z - 2)^2 \leq 49$ which is a circle whose radius is 7. The area is 49π.

19. The points in the intersection of the two spheres satisfy both equations. Rewrite the equations as $(x + 8)^2 + y^2 + (z + 3)^2 - 16 = 0$ and $(x + 8)^2 + (y + 5)^2 + (z + 3)^2 - 9 = 0$ and set them equal to each other. We get $y^2 - 16 = (y + 5)^2 - 9$ or $y = -3.2$. That is, the spheres intersect in the plane $y = -3.2$. In this plane, the equations of the spheres become $(x + 8)^2 + (z + 3)^2 = 16 - (-3.2)^2$, which is the equation of a circle whose center is $(-8, -3.2, -3)$ and whose radius is $\sqrt{5.76}$. The area of this circle is approximately 5.76π.

21. The cylinder tangent to the two coordinate planes is $(x - 1)^2 + (y - 1)^2 = 1$ because its axis is parallel to the z-axis and because its center is at $(1, 1, z)$. The second cylinder has central axis along the line parallel to the z–axis and passing through (3, 1, 0), and so has $(x - 3)^2 + (y - 1)^2 = 1$ as its equation.

23. The intercepts are (0, 0, 6), (0, 3, 0), and (2, 0, 0). The line in the plane $z = 0$ with intercepts (0, 3) and (2, 0) is $2y + 3x = 6$. The line in the yz-plane, where $x = 0$, is $z + 2y = 6$ and the line in the xz-plane, where $y = 0$, is $z + 3x = 6$.

25. The line joining the origin O and the center of the large sphere S passes through the center of the small sphere C and the point T where the two spheres are tangent. Clearly the x, y, and z-coordinates of each of these two points are equal. The coordinates of the origin are (0, 0, 0) and the coordinates of S are (2, 2, 2). Let the coordinates of C be (c, c, c) and those of T be (t, t, t). Since C is c units away from each of the coordinate planes, the equation of a sphere centered at C and tangent to the planes is $(x - c)^2 + (y - c)^2 + (z - c)^2 = c^2$. We need to compute the value of c.

The length of $\overline{CT}$ is c, the radius of the small sphere. We can compute this length, using the distance formula, as
$c = \sqrt{(t - c)^2 + (t - c)^2 + (t - c)^2} = (t - c)\sqrt{3}$.

In order to determine c, we need a second equation involving t.

The length of $\overline{ST}$ is 2, the radius of the large sphere. Therefore, the distance formula gives us
$2 = \sqrt{(2 - t)^2 + (2 - t)^2 + (2 - t)^2} = (2 - t)\sqrt{3}$.

Use this equation to compute $t = 2 - \frac{2}{\sqrt{3}}$ and substitute into the first equation. We get $c = (2 - \frac{2}{\sqrt{3}} - c)\sqrt{3}$. Distribute the multiplication to obtain $c = 2\sqrt{3} - 2 - c\sqrt{3}$ and collect like terms. We get $c + c\sqrt{3} = 2\sqrt{3} - 2$. Next, factor and solve for c: $c(1 + \sqrt{3}) = 2\sqrt{3} - 2$ so that $c = \frac{2\sqrt{3} - 2}{1 + \sqrt{3}}$. By rationalizing the denominator of this number, you can express c more simply as $4 - 2\sqrt{3}$. Thus, the exact value of the radius of the small sphere is $c = 4 - 2\sqrt{3}$.

Chapter 2

1. The x-sections of the plane $z = x + 2y$ are parallel lines with slope 2 and z-intercept x. The y-sections are parallel lines with slope 1 and z-intercept $2y$.

3. a. The x-sections are shown in Figure S2.3a. The x-sections are hyperbolas when $x \neq 0$, and a pair of intersecting lines when $x = 0$. Since x is squared in the equation, the x-sections for $x = a$ and $x = -a$ are identical. Note that the x-sections are curves in a yz-plane.

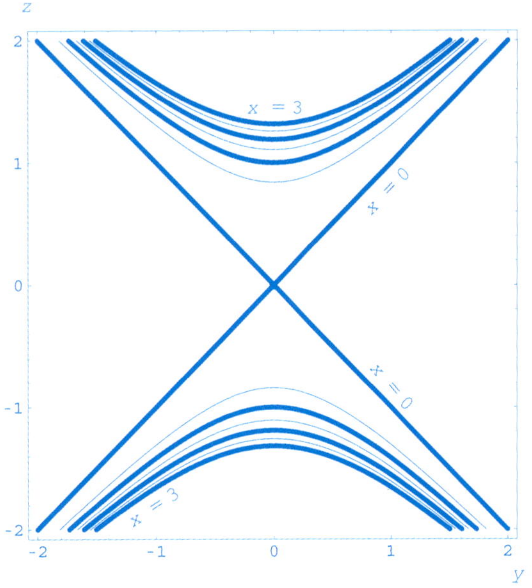

Figure S2.3a

b. The y-sections are identical to the x-sections shown in Figure 2.3a, except that the horizontal axis is the x-axis.

c. The x- and y-sections are hyperbolas and the z-sections are circles. This suggests that the equation is the surface of a pair of cones with common vertex at the origin. Remember that hyperbolas are *conic sections*. That name is to be taken literally. If you slice a cone, you can get a cross-section that is a hyperbola. The surface is shown in Figure S2.3c and a typical x-section is shown in Figure S2.3b.

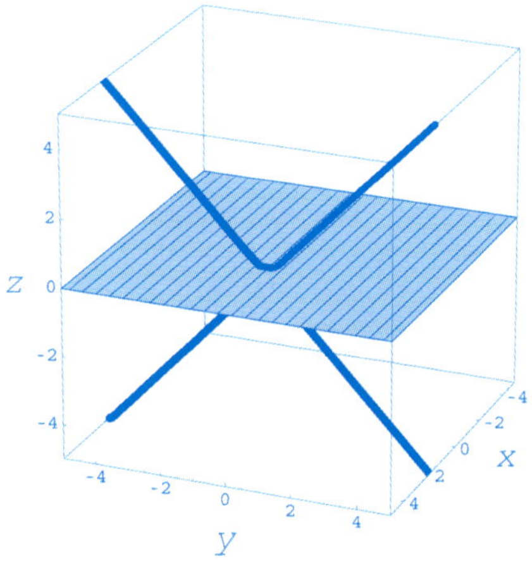

Figure S2.3b

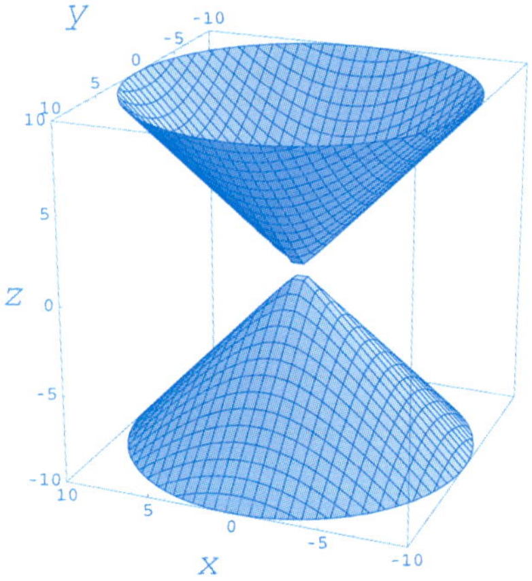

Figure S2.3c

5. The z-sections of both $x^2 + y^2 = z^2$ and $x^2 + y^2 = z$ are concentric circles centered at the origin. The difference is that the radii of the z-sections of $x^2 + y^2 = z^2$ are equal to z, so that the sections are equally spaced. On the other hand, the radii of the z-sections of $x^2 + y^2 = z$ are equal to $\sqrt{z}$, so that these circles grow more and more dense as z increases. Figures S2.5a and S2.5b show these cross-sections.

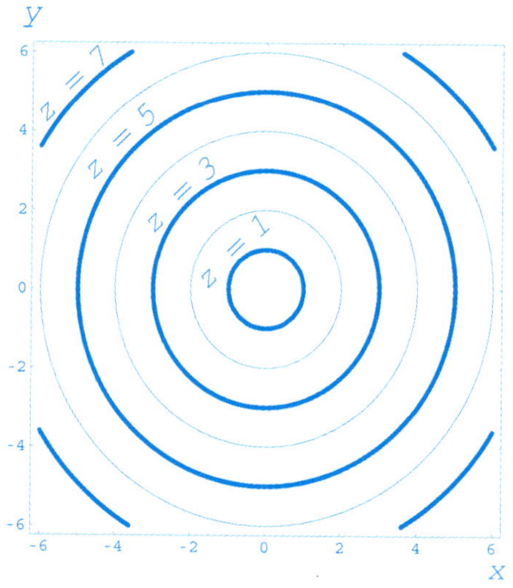

Figure 2.5a

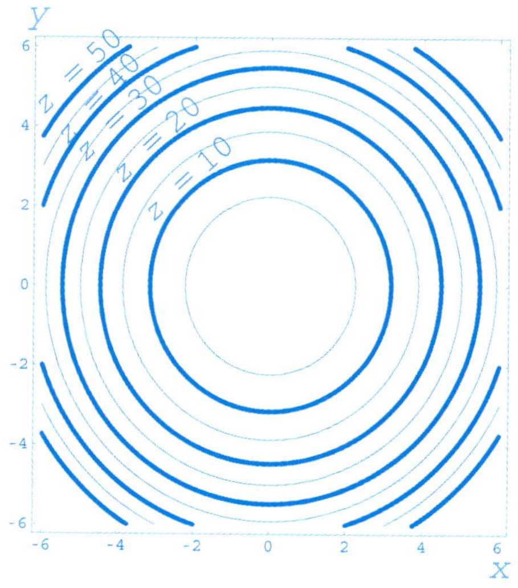

Figure 2.5b

7. When x is held constant, the equation $x^2 - y + z^2 = 0$ will graph as the parabola $y = z^2 + x^2$, so the x-sections are parabolas. These parabolas are not equally spaced, but rather grow more and more sparse as x increases, suggesting that the surface gets less and less steep in the x-direction.

When y is held constant, the equation becomes the equation of a circle, $x^2 + z^2 = y$. These concentric circles are centered at the origin and their radii are $\sqrt{y}$, so they grow more and more dense as y increases. This suggests that the surface rises more and more steeply as we move in the direction of increasing y values. The surface is shown in Figure S2.7.

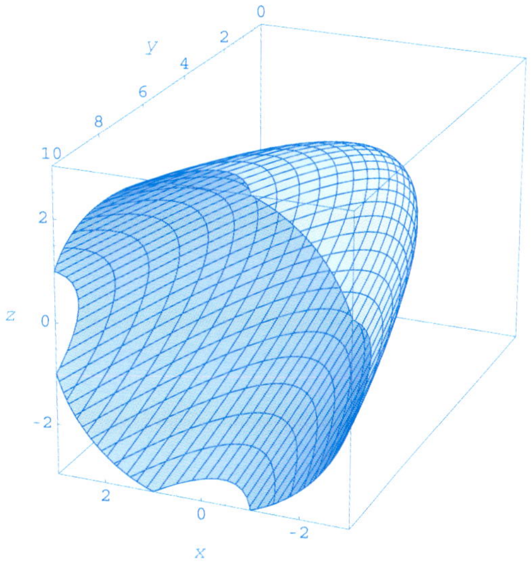

Figure S2.7

9. $(z - y)^2 = x^2 + k$

11. $z = ye^{-x}$, for example. The z-sections are exponential growth functions since, for each value of z, the section is $y = ze^x$.

13. The $x = 0$ section passes through (0, 4) and (2, –2). That is, $y = 0$ when $z = 4$, and $y = 2$ when $z = -2$. Substitute these values into $ax + by + cz = d$ to get $a(0) + b(0) + c(4) = d$ and $a(0) + b(2) + c(-2) = d$. Therefore, $d = 4c$ and so $2b - 2c = d$. Substitute to get $2b - 2c = 4c$ or $b = 3c$.

The $y = 0$ section passes through (2, 0) and (0, 4), so that $2a = d$ and $d = 4c$. Substitute to get $a = 2c$. Substitute the equations for a, b, and d into $ax + by + cz = d$ to get $2cx + 3cy + cz = 4c$. Divide through by the common factor c to get the equation of the plane $2x + 3y + z = 4$.

15. The domain is $L > 0$ and $W > 0$ since the length and width of a rectangle must be positive. An L-section is formed by holding L constant and allowing W and P to vary. The L-sections are half-lines showing the values of W required to produce a perimeter P. Similarly, the W-sections are half-lines showing the values of L required to produce a perimeter P when W is fixed. The P-sections of the plane are line segments with endpoints on the line $P = 2L$ in the plane $W = 0$ and the line $P = 2W$ in the plane $L = 0$. Points on these segments represent possible dimensions of a rectangle with a given perimeter P.

17. Setting $x = 0$ in the general form gives $z = by^2 + ey + f$ and this is equal to $z = -0.5y^2 - 16$, while setting $y = 0$ gives $z = ax^2 + dx + f$ which is equal to $x^2 - 16$. Comparing coefficients we see that $a = 1$, $b = -0.5$, $d = e = 0$, and $f = -16$. To see why $c = 0$, notice that if c were not equal to 0, then each section would be a parabola with vertex off the z-axis. Since all the sections have vertices on the z-axis, c must be zero. Therefore, $z = x^2 - 0.5y^2 - 16$.

Chapter 3

1. Since $a > 0$ and $b < 0$, the slopes of the level curves are negative. Therefore, the graph on the left is the only possibility.

3. All three surfaces have level curves of the form $x + y = k$; however, they differ in the way in which k changes with z. In (i), equal changes in z produce equal changes in k so that Figure 3.12a is the only possibility. In (ii), $k = \sqrt{z}$, and equal changes in z will cause k to increase as the square root of z. Thus, the y-intercepts will increase, but at a slower and slower rate as depicted in Figure 3.12b. In (iii), $k = z^3$, and equal changes in z produce rapidly increasing values of k. Thus, the level curves are increasingly more widely separated as in Figure 3.12c.

5. **a.** $z = -2\sin(\pi x) + y$, for example.
   b. $\frac{1}{z} = -2\sin(\pi x) + y$, for example.

7. Clearly, the function $z = \frac{2 + x + y}{x + y}$ is undefined when $x + y = 0$, which is a plane in space. Therefore, $x + y = 0$ is one of the planes that the surface has as an asymptote. Also note that, as the sum $x + y$ approaches infinity, the function approaches 1. This can be seen by rewriting the fraction as $z = \frac{2}{x + y} + 1$. Therefore, the surface is also asymptotic to the plane $z = 1$.

Both planes are shown in Figure S3.7 on page 86.

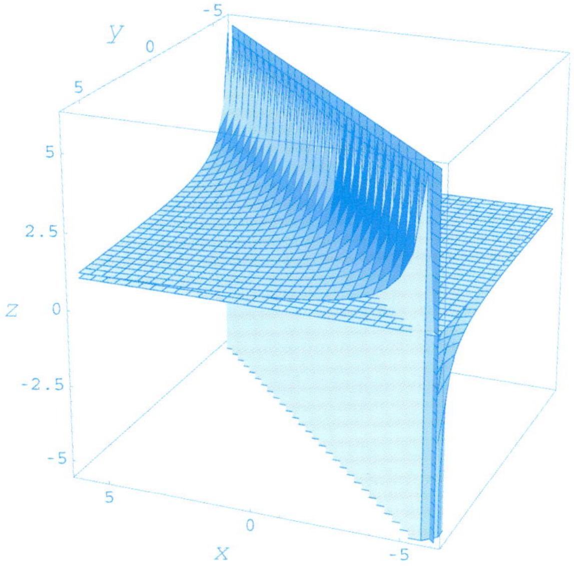

Figure S3.7

9. Consider the level surface $z = 0$ of $z = [x] - [y]$. How can $z = 0$? At first you might think that x and y must be equal, but remember that we are dealing with the greatest integer function. This means that $[x] = 3$, for example, whenever $3 \leq x < 4$. Similarly, $[y] = 3$ whenever $3 \leq y < 4$. Therefore, $[z] = 0$ whenever (x, y) is in the square $[3, 4) \times [3, 4)$. Of course, there was nothing special about the value $[x] = 3$, and the same reasoning would show that if (x, y) is in any unit square whose lower left and upper right corners lie on the line $y = x$ then $[x] = [y]$ and so $z = 0$. So this level of the surface consists of square platforms whose lower left corners have integer coordinates and lie on the line $y = x$.

Similar analysis shows that z is constant on any unit square in the xy-plane. If $n \leq x < n + 1$ and $m \leq y < m + 1$ then $[z] = m + n$. The surface, then, consists of infinitely many unit square "platforms" floating above or below the xy-plane. In the plane $z = c$, for example, lie all square where $[x] - [y] = c$ so if $[x] = n$ then $[y] = c - n$. Points lying in the square $[n, n + 1) \times [c - n, c - n + 1)$ all output platforms at the height $z = c$. These platforms have their lower left corners and upper right corners on the line $y = x - c$.

Since the range of this function consists of integers, there are only z-contours for integer values of z. Each of these consists of infinitely many unit square platforms where the lower left corner of the square lies along a diagonal line $y = x - z$.

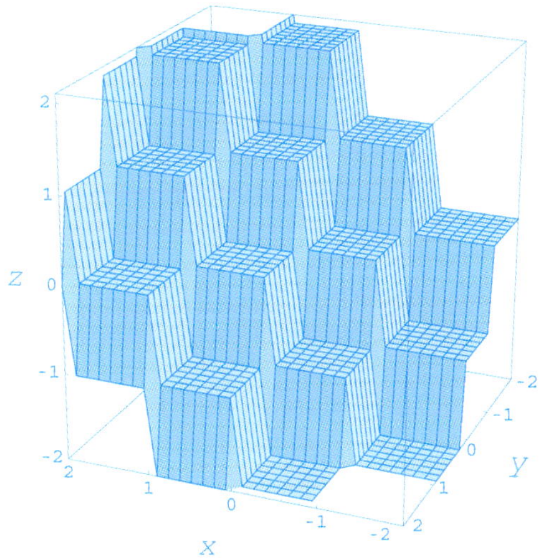

Figure S3.9

Chapter 4

1. a. Do not be misled by the multiplication. Since either x or y is held fixed when differentiating, we do not need the Product Rule. We compute $\frac{\partial z}{\partial x} = y\cos(xy)$ and $\frac{\partial z}{\partial y} = x\cos(xy)$.

Therefore, $\left.\frac{\partial z}{\partial x}\right|_{(\pi, 2, 0)} = 2$ and $\left.\frac{\partial z}{\partial y}\right|_{(\pi, 2, 0)} = \pi$.

b. If there were a vertical tangent plane then one of the partial derivatives would fail to exist at the point of tangency. Since both derivatives are defined for all (x, y), there is no point where the partial derivatives do not exist and, therefore, there is no vertical tangent plane.

3. a. A horizontal line through (2, 2) hits $z = 7$ at (0, 2) and $z = 6$ at (3, 2). Therefore,

$$\left.\frac{\partial z}{\partial x}\right|_{(2,2)} \approx \frac{\Delta z}{\Delta x} \approx \frac{7-6}{0-3} = -\frac{1}{3}.$$

A vertical line through (2, 2) hits $z = 7$ at about (2, 2.8) and $z = 6$ at about (2, 1.8).

Therefore, $\left.\frac{\partial z}{\partial y}\right|_{(2,2)} \approx \frac{\Delta z}{\Delta y} \approx \frac{7-6}{2.8-1.8} = 1.$

b. A horizontal line through (–1, 0) hits $z = 6$ at (–3, 0) and $z = 4$ at (3, 0). Therefore,

$$\left.\frac{\partial z}{\partial x}\right|_{(-1,0)} \approx \frac{\Delta z}{\Delta x} \approx \frac{6-4}{-3-3} = -\frac{1}{3}.$$

A vertical line through the same point hits $z = 6$ at about (–1, 0.8) and $z = 5$ at about (–1, –0.2). Therefore,

$$\left.\frac{\partial z}{\partial y}\right|_{(-1,0)} \approx \frac{\Delta z}{\Delta y} \approx \frac{6-5}{0.8+0.2} = 1.$$

These values are the same as the values in **a.**, as they must be, since the equally spaced lines on the level curves are contours of a plane.

5. The point (–2, 4) lies on the line $y = 4$, which intersects the level curves $z = 2900$ and $z = 2940$ at, approximately, (–2, 4) and (–1.2, 4). These points represent the points (–2, 4, 2900) and (–1.2, 4, 2940) on the surface. Therefore, $f_x(-2,-4) = \frac{2940-2900}{-1.2+2} = \frac{40}{0.8} \approx 50.$
It also lies on the vertical line $x = -2$, which joins, approximately, (–2,–4) and (–2, 1.5). These represent (–2, –4, 2900) and (–2, 1.5, 2940) and so the derivative can be approximated as
$f_y(-2,-4) \approx \frac{2940-2900}{1.5+4} = \frac{40}{5.5} \approx 7.3.$

Similarly, a horizontal line through (4, –2) joins (3.9, –2) on $z = 2980$ with (4.2, –2) on $z = 2960$, while a vertical line through the same point joins (4, –1) on $z = 2980$ and (4, –3) on $z = 2960$. These let us approximate the slopes as $f_x(4,-2) \approx \frac{2960-2980}{4.2-3.9} \approx -67$ and $f_y(4,-2) \approx \frac{2960-2980}{-3+1} \approx 10.$

7. Hold x constant at (1, –1, 3). The nearby values are (1,–2, 7) and (1, 0, 1); these values lie in the same row as $x = 1$. At this point,
$h_y(1,-1) \approx \frac{\Delta z}{\Delta y} = \frac{7-1}{-2-0} = -3.$
Similarly, at (–1, 1, 5), we get
$h_y(1,-1) \approx \frac{\Delta z}{\Delta y} = \frac{7-1}{0-2} = 3.$

9. We use the point-slope formula for a plane,
$z - z_0 = m_x(x - x_0) + m_y(y - y_0),$
where $m_x = \left.\frac{\partial z}{\partial x}\right|_{(x_0,y_0,z_0)}$ and $m_y = \left.\frac{\partial z}{\partial y}\right|_{(x_0,y_0,z_0)}$.
Here we have $\frac{\partial z}{\partial x} = \frac{2x}{(x^2-y^2)}$ and $\frac{\partial z}{\partial y} = \frac{-2y}{(x^2-y^2)}$ so the equation of the tangent plane at (2, 1, ln(3)) is $z = (\frac{4}{3})(x-2) - (\frac{2}{3})(y-1) + \ln(3)$.

11. a. Near (1, –1, 2), x is positive and y is negative, so that the function is $z = x - y$. The partial derivatives are $z_x = 1$ and $z_y = -1$, so that the tangent plane is $z = 1(x-1) - 1(y+1) + 2$ or $z = x - y$.

b. Near (–2, 3, 5), the function is $z = -x + y$, so the partial derivatives are $z_x = -1$ and $z_y = 1$. The tangent plane is $z = -1(x+2) + 1(y-3) + 5$ or $z = -x + y$.

c. If x and y are both positive at some point $(x_0, y_0, |x_0| + |y_0|)$, then the function becomes $z = x + y$ which is the equation of a plane. The equation of a tangent plane is independent of the particular point chosen and is just the surface itself, $z = x + y$.

d. Depending upon the signs of x and y, the function is either the plane $z = x + y$, $z = x - y$, $z = -x + y$, or $z = -x - y$. Therefore, the surface is composed of parts of these four intersecting planes that meet at the origin. Just as $y = |x|$ makes a v-shaped graph in the plane, $z = |x| + |y|$ makes a double v-shaped surface that looks like an infinitely long four-sided pyramid.

13. If $\frac{\partial z}{\partial y} = 4xy - 2y$, then $z = 2xy^2 - y^2 + g(x)$.
Now, differentiate this with respect to x to get $\frac{\partial z}{\partial x} = 2y^2 + g'(x)$. Since we are given that $\frac{\partial z}{\partial x} = 2x + 2y^2$, these two expressions for the derivative must be identical. This means that $g'(x) = 2x$. But this is an ordinary differential equation, so $g(x) = x^2 + C$ and therefore $z = 2xy^2 + x^2 - y^2 + C$.
Since the function passes through (2, 2, 2), we must have $2 = 16 + C$ and so $C = -14$ and the solution is $z = 2xy^2 + x^2 - y^2 - 14$.

15. Begin by antidifferentiating one of the partial derivatives, for example,
$z = \int \frac{\partial z}{\partial x} dx = \int \left(x\sqrt{x^2 + 2y} - 3\right) dx.$

Make the substitution $u = x^2 + 2y$ so $du = 2xdx$ (remember that we are holding y constant when we work with this partial derivative). The antiderivative becomes
$z = \frac{1}{2}\int \sqrt{u}du - \int 3dx = \frac{u^{3/2}}{3} - 3x + f(y)$ or

$z = \frac{(x^2 + 2y)^{3/2}}{3} - 3x + f(y).$

Now, differentiate with respect to y. This gives $\frac{\partial z}{\partial y} = (x^2 + 2y)^{1/2} + f'(y)$. Since we are given that $\frac{\partial z}{\partial y} = \sqrt{x^2 + 2y} + \sin(2y) + 1$, it must be that $f'(y) = \sin(2y) + 1$ and so $f(y) = -\frac{1}{2}\cos(2y) + y + C$ and

$z = \frac{(x^2 + 2y)^{3/2}}{3} - 3x - \frac{1}{2}\cos(2y) + y + C.$

Since $z(-1, 0) = 0$ we must have
$0 = \frac{1}{3} + 3 - \frac{1}{2} + C$ and $C = \frac{17}{6}$. The function is, therefore,

$z = \frac{(x^2 + 2y)^{3/2}}{3} - 3x - \frac{1}{2}\cos(2y) + y + \frac{17}{6}.$

17. $\frac{\partial z}{\partial y} = \lim_{h \to 0} \frac{z(x, y + h) - z(x, y)}{h} =$

$\lim_{h \to 0} \frac{|x| - 2(y + h) - (|x| - 2y)}{h} = \lim_{h \to 0} \frac{-2h}{h} = -2.$

Chapter 5

1. When x is constant, the cross-sections are of the form $y = -z^2 + (16 - x^2)$, which represent ∩-shaped parabolas symmetrical across the z-axis such that increasing x causes the vertex to shift downward. Similarly, when z is constant, the cross-sections are $y = -x^2 + (16 - z^2)$, which are identical parabolas but centered on the x-axis. When y is constant, the cross-sections are concentric circles of the form $x^2 + z^2 = 16 - y$, centered at the origin. When $y = 16$, the cross-section is a point; then, as y decreases, the radius increases.

3.a. When $x = 4$, the cross-section is a circle in a yz-plane whose radius is $y = \frac{1}{16}$. The equation of this circle is $y^2 + z^2 = \frac{1}{256}$.

b. A typical x-section is a circle whose radius is $\frac{1}{x^2}$, so the equation of this section, and of the surface, is $y^2 + z^2 = \frac{1}{x^4}$.

c. When the curve is rotated around the y-axis, a typical y-section is a circle whose radius is the x-value of a point on the curve. This circle has the equation $x^2 + z^2 = r^2$. Since $y = \frac{1}{x^2}$, the square of the radius is just $\frac{1}{y}$. Therefore, the surface is, $x^2 + z^2 = \frac{1}{y}$.

5. **a.** A typical y-section is a circle whose radius is the x-coordinate of a point on the curve. That is, $x = \frac{\ln(y)}{\ln(2)}$. The equation of the surface is, therefore,

$x^2 + y^2 = \left(\frac{\ln(y)}{\ln(2)}\right)^2$.

b. Rotating $z = 2^y$ produces the same surface as in **5a.**, but centered on a different axis. In this case, a typical z-section is a circle whose radius is the y-coordinate of a point on the curve. The equation of the surface is, therefore, $x^2 + y^2 = \left(\frac{\ln(z)}{\ln(2)}\right)^2$.

7. The confusion arises because the x on the left-hand side of $x^2 + z^2 = x^2$ and the x on the right-hand side represent different things. The x on the left-hand side represents any possible value of x on the circular cross-section, and there are infinitely many such values. The x on the right-hand side represents the particular radius of the cross-section, and there is only one such value for each cross-section. To make the equation meaningful, we must replace the x on the right-hand side with a formula that computes the particular x-value.

9. The $z = 0$ contour is the parabola $y = x^2 - 4$; this is the curve that is rotated around the y-axis. A typical disc has radius x, so the volume of a typical disc is $\Delta V = \pi x^2 \Delta y$ or $\Delta V = \pi(y + 4)\,\Delta y$. The volume of this solid of revolution is,

therefore, $V = \pi\int_{-4}^{0}(y + 4)dy = 8\pi$.

11. **a.** A typical y-section of the surface is a circle whose radius is an x-value on the curve, so the equation of a typical y-section is $x^2 + z^2 = y^2 - 1$. The volume of a typical disc is $\Delta V = \pi(y^2 - 1)\Delta y$, leading

to the integral $V = \pi\int_{1}^{3}(y^2 - 1)dy = \frac{20\pi}{3}$.

b. The hyperbola is asymptotic to the lines $y = x$ and $y = -x$. These lines rotate into the surface of the cones $x^2 + z^2 = y^2$. Each of the sheets of the hyperboloid lies inside one of these opposite facing cones and approaches it asymptotically.

13. This surface is the same as the surface $z = x^2 + y^2$ shifted 2 units in the positive x-direction and 4 units in the positive y-direction. Since the paraboloid $z = x^2 + y^2$ is formed by rotating $z = x^2$ around the z-axis, the paraboloid $z = (x - 2)^2 + (y - 4)^2$ is formed by rotating $z = (x - 2)^2$ around the line perpendicular to the xy-plane at (2, 4) because the axis shifts over 2 and up 4 as well.

A typical disc has radius $\sqrt{z}$ and thickness Δz so the volume of this solid is $V = \pi\int_{0}^{9} z\,dz = 40.5\pi$.

15. The two surfaces intersect in a circle and the volume between them is the difference of the volumes between the paraboloids and the plane in which they intersect. To find this plane, set the two equations equal to each other and solve. This gives $2x^2 + 2y^2 + 26 = \frac{1}{2}x^2 + \frac{1}{2}y^2 + 38$, which reduces to $x^2 + y^2 = 8$. That is, the surfaces intersect in a circle whose radius is $\sqrt{8}$. Substitute $x^2 + y^2 = 8$ back in one of the original equations to get that $z = 42$. Therefore, the two surfaces intersect in the plane $z = 42$. Since their vertices are (0, 0, 26) and (0, 0, 38), the volume between the surfaces can be computed as the difference between the volumes of two paraboloids of revolution and $V = \pi\int_{26}^{42}\frac{z - 26}{2}dz - \pi\int_{38}^{42}2(z - 38)dz = 48\pi$.

17. a. A typical x-section of the horn has radius y so that $y^2 + z^2 = \frac{1}{x^2}$.

b. The volume is given by the improper integral $\pi\int_1^{\infty} \frac{1}{x^2}dx$. If the volume exists it can be computed from

$$\lim_{b\to\infty} \pi\int_1^b \frac{1}{x^2}dx = \lim_{b\to\infty}\left(-\frac{\pi}{x}\right)\Big|_1^b = \lim_{b\to\infty}\left(-\frac{\pi}{b} + \pi\right) = \pi.$$

Since the limit exists, the integral converges and the volume is $V = \pi$.

Chapter 6

1. $\int_0^{\pi/2} 3\sin(x + z)dx = -3\cos(x + z)\Big|_0^{\pi/2}$

$$= -3\cos\left(\frac{\pi}{2} + z\right) + 3\cos(z) \text{ so}$$

$$\int_0^{\pi}\int_0^{\pi/2} 3\sin(x + z)dxdz$$

$$= \int_0^{\pi}\left(-3\cos\left(\frac{\pi}{2} + z\right) + 3\cos(z)\right)dz$$

$$= \left(-3\sin\left(\frac{\pi}{2} + z\right) + 3\sin(z)\right)\Big|_0^{\pi}$$

$$= -3\sin\left(\frac{3\pi}{2}\right) + 3\sin(\pi) + 3\sin\left(\frac{\pi}{2}\right) - 3\sin(0) = 6.$$

3. $V = \int_{-\pi/2}^{\pi/2}\int_{-\pi/2}^{\pi/2}(\cos(x) + \cos(y))dxdy.$

The inner integral is

$$\int_{-\pi/2}^{\pi/2}(\cos(x) + \cos(y))dx$$

$$= (\sin(x) + x\cos(y))\Big|_{-\pi/2}^{\pi/2} = 2 + \pi\cos(y)$$

so $V = \int_{-\pi/2}^{\pi/2}\int_{-\pi/2}^{\pi/2}(\cos(x) + \cos(y))dxdy$

$$= \int_{-\pi/2}^{\pi/2}(2 + \pi\cos(y))dy$$

$$= (2y + \pi\sin(y))\Big|_{-\pi/2}^{\pi/2}$$

$$= 4\pi.$$

5. $V = \int_0^4\int_0^2 e^{x+y}dxdy = \int_0^4 e^{x+y}\Big|_{x=0}^{x=2}dy$

$= \int_0^4 (e^{2+y} - e^y)dy = e^6 - e^4 - e^2 + 1.$

7. The volume of the hemisphere is given by the double integral

$$V = \iint_A \sqrt{r^2 - x^2 - y^2}\,dA,$$

which we will evaluate as an iterated integral. Here A stands for the base of the hemisphere, the circular region $x^2 + y^2 \le r^2$ in the xy-plane.

Suppose we slice perpendicular to the y-axis and then add up the volumes of these slabs.

The volume will be $\int_{-r}^{r} A(y)dy$ where A gives the area of a typical y-section.

That is, the volume is the iterated integral

$$V = \int_{-r}^{r}\int_{-r}^{r}\sqrt{r^2 - x^2 - y^2}\,dydx.$$

An integral such as this can be evaluated algebraically, although the solution is messy. Recall that the trick to evaluating an integral with a square root is to find a substitution that transforms the expression under the radical into a perfect square. If this were the single variable integral $\int_{-r}^{r}\sqrt{r^2 - x^2}\,dx$, the trigonometric substitution $x = r\sin\theta$ would do the trick since $r^2 - x^2 = r^2(1 - \sin^2\theta)$. However, it is much easier to solve this integral geometrically, by recognizing the integral as the area of a semi-circle of radius r.

Our multivariable integral can also be evaluated in two ways. The substitution $y = \sqrt{r^2 - x^2}\sin\theta$ would transform the integrand into

$$\sqrt{r^2 - x^2 - y^2}$$

$$= \sqrt{r^2 - x^2 - (r^2 - x^2)\sin^2\theta}$$

$$= \sqrt{(r^2 - x^2)(1 - \sin^2\theta)} = \sqrt{r^2 - x^2}\cos\theta.$$

Since $dy = \sqrt{r^2 - x^2}\cos\theta d\theta$, the integral becomes

$$\int_0^{\pi}(r^2 - x^2)\cos^2\theta d\theta = (r^2 - x^2)\left(\frac{\theta}{2} + \frac{\sin 2\theta}{4\theta}\right)\Big|_0^{\pi}$$
$$= \frac{\pi}{2}(r^2 - x^2).$$

However, the geometric solution is much easier. Notice that $\int_{-r}^{r}\sqrt{r^2 - x^2 - y^2}dy$ can be interpreted as the area of a semi-circle whose radius is $y = \sqrt{r^2 - x^2}$ and is, thus, $\frac{\pi(r^2 - x^2)}{2}$. The volume is, therefore,

$$V = \frac{1}{2}\int_{-r}^{r}\pi(r^2 - x^2)dx = \frac{\pi}{2}\left(r^2x - \frac{1}{3}x^3\right)\Big|_{-r}^{r}$$
$$= \frac{\pi}{2}\left(\frac{4}{3}r^2\right) = \frac{2}{3}\pi r^3.$$

Chapter 7

1. The average is

$$\frac{1}{(\frac{\pi}{2})^2}\iint_{[0,\pi/2]\times[0,\pi/2]}(x\sin(y) + y\sin(x))dA$$
$$= \frac{4}{\pi^2}\int_0^{\pi/2}\int_0^{\pi/2}(x\sin(y) + y\sin(x))dxdy.$$

Begin by evaluating the inner integral.

$$\int_0^{\pi/2}(x\sin(y) + y\sin(x))dx$$
$$= \left(\frac{x^2\sin(y)}{2} - y\cos(x)\right)\Big|_0^{\pi\backslash 2} = \frac{\pi^2}{8}\sin(y) + y.$$

Therefore, the average value is

$$\frac{4}{\pi^2}\int_0^{\pi/2}\left(\frac{\pi^2}{8}\sin(y) + y\right)dy = \left(-\frac{1}{2}\cos(y) + \frac{2y^2}{\pi^2}\right)\Big|_0^{\pi\backslash 2} = 1.$$

3. Average $= \iint_{[1,2]\times[1,2]}\frac{x^2 + y^2}{xy}dA = \int_1^2\int_1^2\left(\frac{x}{y} + \frac{y}{x}\right)dxdy.$

The inner integral is

$$\int_1^2\left(\frac{x}{y} + \frac{y}{x}\right)dx = \left(\frac{x^2}{2y} + y\ln|x|\right)\Big|_1^2 = \frac{3}{2y} + y\ln(2).$$

The average value is, therefore,

$$= \int_1^2\left(\frac{3}{2y} + y\ln(2)\right)dy = \left(\frac{3}{2}\ln|y| + \frac{y^2\ln(2)}{2}\right)\Big|_1^2$$
$$= \frac{3}{2}\ln(2) + 2\ln(2) - \frac{1}{2}\ln(2) = 3\ln(2) \approx 2.079.$$

5. a. The plane $z = 12 - 2x$ intersects the plane $x = 0$ at $z = 12$ and the plane $x = 2$ at $z = 8$. Therefore, the coordinates of the slant top of the box are (0, 0, 12), (0, 3, 12), (2, 0, 8) and (2, 3, 8). The slanted box can, thus, be viewed as a rectangular box topped by a right triangular prism as shown in Figure S7.5a.

Figure S7.5a

The box has a 2×3 rectangular base and a height of 8. The triangular prism has the same rectangular base with a height of $12 - 8 = 4$.

The average height the box is 8 because every height within the box is 8. The average height of the prism is the volume of the prism divided by the area of the base or $\frac{(1/2)(2)(4)(3)}{(2)(3)} = 2$. The average height of the slanted box is, therefore, the sum of these heights is $8 + 2 = 10$.

b. The average height of the box is given by the double integral $\frac{1}{6}\iint_{[0,2]\times[0,3]} z\,dA$. Write this as Average $= \frac{1}{6}\int_0^3\int_0^2 (12 - 2x)dxdy$.

The inner integral is $\int_0^2 (12 - 2x)dx = 20$.

Therefore, the average height of the slant-top box is $\frac{1}{6}\int_0^3 20dy = 10$.

7. The cost function $C = 0.2x^2 - 0.003xy + 0.1y^2 + 18$ is evaluated over the rectangle $500 \le x \le 1000$ and $300 \le y \le 700$. This is a rectangle whose base has area $(500)(400) = 200{,}000$. The average value is the double integral

Average =

$$\frac{1}{200{,}000}\iint_{[500,1000]\times[300,700]} (0.2x^2 - 0.003xy + 0.1y^2 + 18)dA,$$

which we compute as the iterated integral

$$\frac{1}{200{,}000}\int_{500}^{1000}\int_{300}^{700} (0.2x^2 - 0.003xy + 0.1y^2 + 18)dydx.$$

The inner integral is

$$\int_{300}^{700} (0.2x^2 - 0.003xy + 0.1y^2 + 18)dy =$$

$$\left(0.2yx^2 - \frac{.003}{2}xy^2 + \frac{0.1}{3}y^3 + 18y\right)\Big|_{300}^{700}$$

$$= 0.2(400)x^2 - \frac{0.003x}{2}(400{,}000) + \frac{0.1}{3}(316{,}000{,}000) +$$

$$7{,}200 = 80x^2 - 600x + \frac{31{,}621{,}600}{3}.$$

And so, the average value is

$$\frac{1}{200{,}000}\int_{500}^{1000}\left(80x^2 - 600x + \frac{31{,}621{,}600}{3}\right)dx$$

$$= \frac{1}{200{,}000}\left(\frac{80}{3}x^3 - 300x^2 + \frac{31{,}621{,}600x}{3}\right)\Big|_{500}^{1000}$$

$$\approx \frac{1}{200{,}000}\left(\frac{80}{3}(875{,}000{,}000) - 300(750{,}000) + \right.$$

$$\left. 5{,}270{,}266{,}667\right) = 141{,}893$$

Therefore, the average cost is \$141,893.

9. The base of the box is a 2×3 rectangle, so the average height of the box is

Average $= \frac{1}{6}\iint_{[0,2]\times[0,3]} z\,dA$.

We can compute the double integral as an iterated integral in two ways, and, as we have seen, we will get the same answer whichever way we use. Let's compute the average by first slicing in the y-direction.

This gives Average $= \frac{1}{6}\int_0^3\int_0^2 \left(9 - \frac{x}{2} - y\right)dxdy$

and the inner integral is

$$\int_0^2 \left(9 - \frac{x}{2} - y\right)dx = \left(9x - \frac{x^2}{4} - xy\right)\Big|_0^2 = 17 - 2y.$$

The average is, therefore,

$$\frac{1}{6}\int_0^3 (17 - 2y)dy = \frac{1}{6}(17y - y^2)\Big|_0^3 = 7.$$

11. a. $\int_0^4\int_{2y}^3 (x + y)dx\,dy = \int_0^4 \left(\frac{x^2}{2} + xy\right)\Big|_{2y}^3 dy$

$$= \int_0^4 \left(\frac{9}{2} + 3y - 2y^2 - 2y^2\right)dy$$

$$= \left(\frac{9y}{2} + \frac{3y^2}{2} - \frac{4y^3}{3}\right)\Big|_0^4 = -\frac{130}{3} \approx -43.333.$$

b. $\int_0^{\sqrt{\frac{\pi}{2}}}\int_0^x x^2\cos(xy)dy\,dx = \int_0^{\sqrt{\frac{\pi}{2}}} x\sin(xy)\Big|_0^x dx$

$= \int_0^{\sqrt{\frac{\pi}{2}}} x\sin(x^2)dx$

$= -\frac{1}{2}\cos(x^2)\Big|_0^{\sqrt{\frac{\pi}{2}}} = -\frac{1}{2}\cos(\frac{\pi}{2}) + \frac{1}{2} = \frac{1}{2}$

c. $\int_0^1\int_{-x}^x x^2y^3dy\,dx = \int_0^1 \frac{x^2y^4}{4}\Big|_{-x}^x dx$

$= \int_0^1\left(\frac{x^6}{4} - \frac{x^6}{4}\right)dx = 0$

13. Whether we slice in the x-direction first or in the y-direction first, a typical slice will extend from one side of the unit circle to the other. If we slice as shown in Figure S7.13, the outer integral will add up these slices and so will be a 'dx' integral, while the inner integral will compute the area of each section with limits of integration from the bottom to the top of the unit circle.

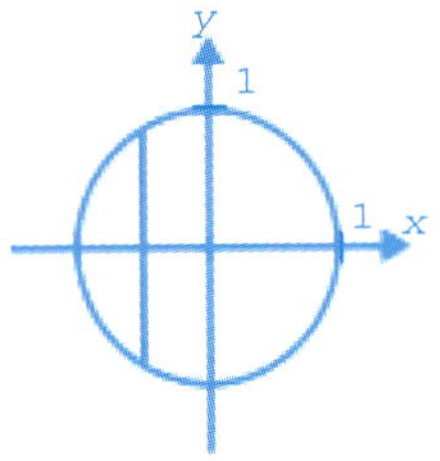

Figure S7.13

That is, $\int_{-1}^1\int_{-\sqrt{1-x^2}}^{\sqrt{1-x^2}}(3x + 2y + 4)dy\,dx$. The inner integral evaluates to

$\int_{-\sqrt{1-x^2}}^{\sqrt{1-x^2}}(3x + 2y + 4)dy$

$= (3xy + y^2 + 4y)\Big|_{\sqrt{-1-x^2}}^{\sqrt{1-x^2}}$

$= 6x\sqrt{1-x^2} + 8\sqrt{1-x^2}$, and so the volume is $\int_{-1}^1(6x\sqrt{1-x^2} + 8\sqrt{1-x^2})dx = 4\pi$

Chapter 8

1. The top of the box is $z = 9 - \frac{x}{2} - y$ and the bottom of the box lies in the xy-plane, so points inside the box must satisfy $0 < z < 9 - \frac{x}{2} - y$. The sides of the box are the planes $x = 0$, $x = 2$, $y = 0$, and $y = 3$, so the interior points must also satisfy $0 < x < 2$ and $0 < y < 3$.

3. Since the sides of the box lie in the planes $x = -3$, $x = 11$, $y = -5$, $y = -1$, $z = -2$, and $z = 14$, the corners are $(-3, -5, -2)$, $(-3, -5, 14)$, $(-3, -1, -2)$, $(-3, -1, 14)$, $(11, -5, -2)$, $(11, -5, 14)$, $(11, -1, -2)$, and $(11, -1, 14)$.

5. a. Figure S8.5a shows the x-sections.

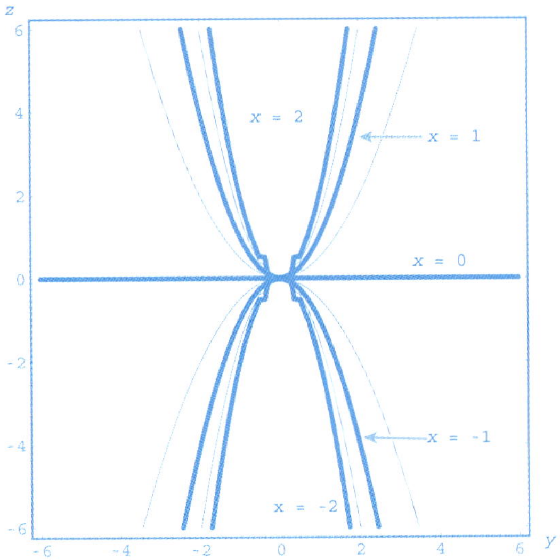

Figure S8.5a

These are parabolas of the form $z = ky^2$, where k represents the (constant) value of x on a particular section. These x-values act as a vertical stretch, so that as x increases the parabolas grow faster. The vertex of each section graph appears at the origin. In space, however, these vertices lie along points $(x, 0, 0)$, that is, along the x-axis, as we can see in Figure S8.5c.

b. The y-sections are lines of the form $z = kx$, where k represents the (constant) value of y on particular sections. These lines all pass through the origin and grow steeper as $|y|$ increases. Figure S8.5b shows these sections. The surface is shown in Figure S8.5c.

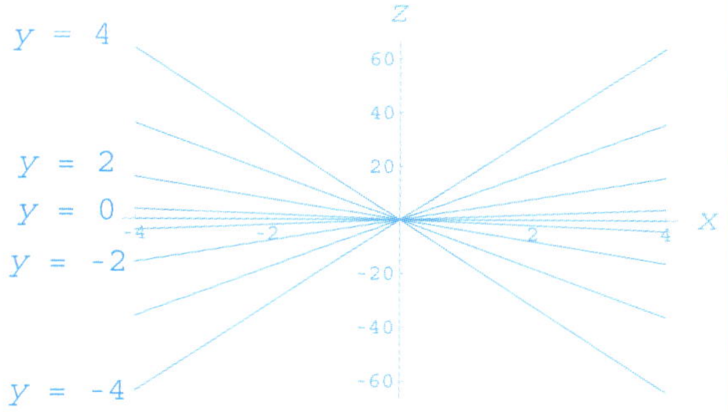

Figure S8.5b

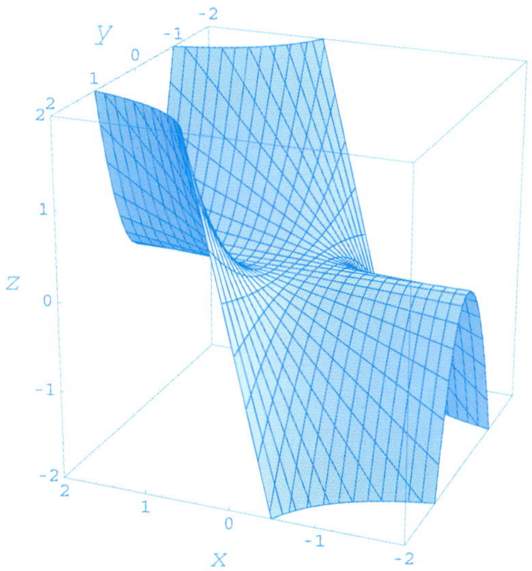

Figure S8.5c

7. The y-sections appear to be parabolas, while the x-sections appear to be sinusoidal waves. The actual equation of the surface is $z = y^2 + \sin(x)$.

9. Since the ellipses are parallel to $x = 0$, any elliptical equation in y and z will do. Specifically, since the x-contours are not specified, any function whose form is $f(x) = \frac{y^2}{a^2} + \frac{z^2}{b^2}$ where a and b are constants is fine. If $a = 1$ and $b = \frac{1}{2}$, for example, we get $f(x) = y^2 + 4z^2$.

11. If $\frac{\partial z}{\partial x} = -3\sin(x)\sin(y)$, then we can antidifferentiate with respect to x to get

$$z = \int -3\sin(x)\sin(y)dx$$
$$= 3\sin(y)\cos(x) + g(y).$$

Now differentiate with respect to y to get

$$\frac{\partial z}{\partial x} = 3\cos(x)\cos(y) + g'(y) \text{so } g'(y) = 0$$

and g is constant. This means that $z = 3\sin(y)\cos(x) + C$. Since we are told that the surface passes through the origin, the constant is zero and so $z = 3\sin(y)\cos(x)$.

13. a. A typical y-section is a circle whose radius is the x-value at that height on $y = \ln(x)$. That is, the radius is e^y. The equation of the surface is, therefore, $x^2 + z^2 = e^{2y}$.

b. The volume can be computed using the method of discs.

$$V = \pi\int_0^{\ln(2)} e^{2y}dy = \frac{3\pi}{2}.$$

c. The unbounded volume is given by $V = \pi\int_{-\infty}^{\ln(2)} e^{2y}dy$. This integral is bounded and can be computed the same way the integral in **b.** was computed. This gives

$$V = \lim_{b\to-\infty}\left[\frac{1}{2}e^{2y}\right]_b^{\ln(2)} = \lim_{b\to-\infty}\frac{\pi}{2}[e^{2\ln(2)} - e^b] = 2\pi.$$

15. a. When $y = x^{1/3} - 1$ is rotated around the y-axis, a typical y-slice is a circle whose radius is x. Solve for x to get $x = (y + 1)^3$. The surface is, therefore, $x^2 + z^2 = (y + 1)^6$.

b. When the equation is rotated around the x-axis, a typical x-slice is a circle whose radius is y. The surface is, therefore, $y^2 + z^2 = (x^{1/3} - 1)^2$.

c. The volume between the planes $|x| = 8$ is $V = \pi\int_{-8}^{8}(x^{1/3} - 1)^2 dx = 54.4\pi$.

17. a. $V = \iint_{[-4,4]\times[-4,4]}(x^3 + y^3)dA$ may seem like the correct integral for computing the volume, but as we will see in part **b.**, it is not. The explanation of how to compute the correct volume is given in part **c.**

b. The function $z = x^3 + y^3$ is an odd function. Since the region of integration is symmetrical around the origin, any tiny region $\Delta x\Delta y$ over which z is positive is exactly balanced by a region of the same size over which z is negative, producing an approximating sum of zero. We saw the same phenomenon in Chapter 6 Problem 4. We can confirm this by direct integration. The inner integral is

$$\int_{-4}^{4}(x^3 + y^3)dx = \left(\frac{x^4}{4} + xy^3\right)\Big|_{-4}^{4} = 8y^3,$$

and so $V = \int_{-4}^{4}8y^3dy = \frac{8y^4}{4}\Big|_{-4}^{4} = 0$.

c. To compute the correct volume, we need to compute the positive volume and then double it. To begin, note that $x^3 + y^3$ is always positive in the first quadrant, and always negative in the third quadrant. In fact, the positive volume above the first quadrant exactly balances the negative volume below the third quadrant. Therefore, if we compute the volume above the first quadrant and double it, we will have determined the total volume between the surface and the first and third quadrants. The volume above the first quadrant can be computed as follows:

$$\int_0^4\int_0^4(x^3 + y^3)dxdy = \int_0^4\left[\frac{x^4}{4} + xy^3\right]_0^4 dy$$

$$= \int_0^4(64 + 4y^3)dy = 512.$$

Therefore, the total volume above the first quadrant and below the third quadrant is $2(512) = 1{,}024$.

The situation in the second and fourth quadrants is trickier, because, in these quadrants, $x^3 + y^3$ is sometimes positive and sometimes negative. Let's focus first on the second quadrant. In this quadrant, we can guarantee that $x^3 + y^3 > 0$ by keeping $y > |x|$. This is true in half of the region in the second quadrant, specifically in the right triangular region whose corners are (0, 0), (–4, 0), and (0, 4). Thus, if we integrate above this region, we will obtain the positive volume above the second quadrant.

$$\int_0^4\int_{-y}^0(x^3 + y^3)dxdy = \int_0^4\left[\frac{x^4}{4} + xy^3\right]_{-y}^0 dy$$

$$= \int_0^4\frac{3}{4}y^4dy = \frac{3}{5}\times 256 = 153.6.$$

Next, we need to double this volume to balance the negative volume in the second quadrant where $y < |x|$. Therefore, the total volume enclosed above and below the second quadrant is $2(153.6) = 307.2$.

The situation in the fourth quadrant is the same, and thus the total volume enclosed by the fourth quadrant is also 307.2.

Overall, the total volume enclosed by the function $z = x^3 + y^3$ is $1024 + 307.2 + 307.2 = 1638.4$

19. The volume above the circle $x^2 + y^2 = 4$ can be found by the method of slabs. Slices parallel to the xz-plane range from $y = -2$ to $y = 2$ and the base of the solid has both ends on opposite sides of the circle, that is, from to $x = -\sqrt{4 - y^2}$ to $x = \sqrt{4 - y^2}$. The iterated integral that gives the volume is, therefore,

$$V = \int_{-2}^{2}\int_{-\sqrt{4-y^2}}^{\sqrt{4-y^2}}(y^4 + x^2)dxdy.$$

The inner integral is

$$\int_{-\sqrt{4-y^2}}^{\sqrt{4-y^2}} (y^4 + x^2)dx = (y^4x + \tfrac{1}{3}x^3)\Big|_{-\sqrt{4-y^2}}^{\sqrt{4-y^2}}$$

$$= 2y^4\sqrt{4-y^2} + \tfrac{2}{3}(4-y^2)^{3/2}.$$

The volume is, therefore,

$$V = \int_{-2}^{2} (2y^4\sqrt{4-y^2} + \tfrac{2}{3}(4-y^2)^{3/2})dy$$

which can be approximated numerically as $V = 37.699$.

Self-Assessment

1. Solve the equations simultaneously to get $y^2 = 11$ so that $y = \pm\sqrt{11}$. Therefore, any point (x, y, z) must have the form $(x, \pm\sqrt{11}, z)$ where $x^2 + z^2 = 25$. For example, $(3, \sqrt{11}, 4)$, $(-1, -\sqrt{11}, \sqrt{24})$, or $(-5, \sqrt{11}, 0)$.

3. The sections let us read the slope in the x-direction as $-\frac{2}{3}$, and the slope in the y- direction as –2. They also show us that the point (0, 1, 0) is on the surface. The surface is a plane, and the equation of the plane is, therefore, $z = \left(-\frac{2}{3}\right)x - 2(y-1)$. The plane intercepts the axes at (3, 0, 0), (0, 1, 0), and (0, 0, 2). A quick sketch clearly shows that the plane is between the bug at (2, 3, 6) and the origin, and so would prevent the bug from reaching it.

5. a. Horizontal lines move from contours with larger z-values to contours with smaller z-values in the second quadrant, and from smaller z-values to larger z-values in the fourth quadrant. In the first and third quadrants, horizontal lines sometimes move to larger and sometimes to smaller z-values. Therefore, $\frac{\partial z}{\partial x} > 0$ in the fourth quadrant and $\frac{\partial z}{\partial x} < 0$ in the second quadrant.

   b. Vertical lines move from contours with smaller z-values to contours with larger z-values in the second quadrant, and from larger z-values to smaller z-values in the fourth quadrant. In the first and third quadrants vertical lines sometimes move to larger and sometimes smaller z-values. Therefore, $\frac{\partial z}{\partial y} > 0$ in the second quadrant and $\frac{\partial z}{\partial y} < 0$ in the fourth quadrant.

7. a. A typical y-slice is a circle whose radius is an x-value of the function $y = \sqrt{x}$, that is, the radius is y^2. The surface is, therefore, $x^2 + z^2 = y^4$.

   b. A typical x-slice is a circle in an yz-plane with radius $y = \sqrt{x}$, so the equation of the surface is $y^2 + z^2 = x$.

9. The paraboloids are centered on the axis passing through (1, 5, 0) and parallel to the z-axis. The volume of each paraboloid can be computed using the method of discs. To find the limits of integration, we set the two functions equal to each other to find their common height. If $12 - (x-1)^2 - (y-5)^2$ $= \frac{1}{4}[(x-1)^2 + (y-5)^2] - 8$ then $(x-1)^2 + (y-5)^2 = 16$, so the paraboloids intersect in a circle of radius 4. Substitute this equation back into either of the original functions to get $z = 12 - 16 = -4$, the level at which the two paraboloids intersect. Since the high point of $z = 12 - (x-1)^2 - (y-5)^2$ lies in $z = 12$ and the low point of $z = \frac{1}{4}[(x-1)^2 + (y-5)^2] - 8$ lies in $z = -8$, the volume enclosed inside the paraboloids is

$$V = \pi\int_{-8}^{-4} 4(z+8)dz + \pi\int_{-4}^{12}(12-z)dz. \text{ So,}$$

$$V = 4\pi\left(\tfrac{1}{2}z^2 + 8z\right)\Big|_{-8}^{-4} + \pi\left(12z - \tfrac{1}{2}z^2\right)\Big|_{-4}^{12}$$

$$= 4\pi(8) + \pi(128) = 160\pi.$$

Notes

Notes